JAPAN'S ECONOMIC AID

Japan's Economic Aid

POLICY-MAKING AND POLITICS

ALAN RIX

ST. MARTIN'S PRESS NEW YORK

St. Martin's Press, Inc., 175 Fifth Avenue, New York, N.Y. 10010
Printed in Great Britain
First published in the United States of America in 1980
ISBN 0-312-44063-4

Library of Congress Cataloging in Publication Data
Rix, Alan.
    Japan's economic aid.
    Bibliography: p. 274
    Includes index.
    1. Economic assistance, Japanese.
2. Technical assistance, Japanese. 3. Japan —
Economic policy. I. Title.
HC60.R55   338.91'51'01724   80-15431

CONTENTS

List of Tables and Figures

Glossary

Preface

Introduction   11

PART I: AID IDEAS AND AID STRUCTURES   19

1. Foreign Aid and the Ministries   21

2. 'Scrap and Build': The Origins of JICA   49

PART II: THE DOMESTIC POLITICS OF FOREIGN AID   81

3. Aid and the Governmental Process   83

4. Ministries and the Policy Process   118

5. Budgeting for Foreign Aid   150

PART III: THE POLITICS OF AID RELATIONSHIPS   189

6. Projects, Surveys and Consultants   191

7. Managing Bilateral Relations   221

8. Policy at Work: The Cycle of Aid   250

Conclusion   267

Select Bibliography   274

Index   278

TABLES AND FIGURES

Tables

1.1 The Net Flow of Financial Resources from Japan to Developing Countries and Multilateral Agencies, 1960-78 ($ million)   32

1.2 Geographical Distribution of Net Flows of Financial Resources from Japan to Developing Countries (percentage)   34

3.1 Staff of Aid Divisions in Main Ministries, 1965-75   93

3.2 Foreign Ministry Economic Cooperation Bureau: Actual Staff Numbers as of 1 April 1976   95

5.1 Economic Cooperation Budget, According to DAC Categories, 1965-77 (Fiscal Year, Yen Million)   154

5.2 Shares of Various Policy Categories in the Budget's General Account, 1965-77 (Fiscal Year, Thousand Million Yen)   156

5.3 Economic Cooperation Budget, General Account: Changes in Ministry Shares, 1965-77   158

5.4 Total Budget Request and Economic Cooperation Budget Request per Ministry, 1975-7: Request Divided by Previous Year's Allocation (Fiscal Year, Percentage)   170

5.5 Economic Cooperation Budget, 1975 and 1976: Sizes of Allocations and Requests (Fiscal Year, Percentage)   171

5.6 Economic Cooperation Budget, 1975 and 1976: Allocation Divided by Previous Year's Allocation (Fiscal Year)   173

5.7 Budget General Account: MOF Draft and Government Draft Allocations for Various Categories, 1965-77 (Fiscal Year, Hundred Million Yen)   178

5.8 Budget Allocations for OECF and Export-Import Bank, 1965-77: MOF Draft and Government Draft (Hundred Million Yen)   180

6.1 Trends in Development Survey Budget, 1962-76 (Million Yen)   194

6.2 Number of Development Survey Teams Sent, 1962-76, by Type   195

6.3	Number of Development Survey Teams Sent, 1962-76, by Region	196
6.4	Geographical Distribution of Development Survey Expenditure, 1954-76 (Million Yen)	197
7.1	Geographical Distribution of Japanese Government Loans: Accumulated Value of Commitments up to Selected Years (Hundred Million Yen)	224
7.2	The Ten Main Recipients of Japanese Government Loans: Accumulated Value of Commitments up to Selected Years (Hundred Million Yen)	225

Figures

3.1	Japanese Aid Administration: Formal Outline	86
3.2	The Structure of Japan's Aid Administration, 1978: The Main Ministries	88
4.1	Economic Cooperation: Structure and Process	120
5.1	Sources of Budget Funds for Official Aid	152

GLOSSARY

Throughout this book, Japanese names are given in the Japanese style with surname first. This is done also for Japanese authors of English language works.

ADB	Asian Development Bank
AMA	Administrative Management Agency
ASEAN	Association of Southeast Asian Nations
DAC	Development Assistance Committee of the Organisation for Economic Cooperation and Development
DAG	Development Assistance Group
EPA	Economic Planning Agency
Eximbank	Export-Import Bank of Japan
GNP	Gross National Product
IGGI	Inter-Governmental Group on Indonesia
JEMIS	Japan Emigration Service
JICA	Japan International Cooperation Agency
Keidanren	*Keizai dantai rengōkai* or Federation of Economic Organisations
Keizai Dōyūkai	Japan Committee for Economic Development
LDC	less developed country
LDP	Liberal Democratic Party
LLDC	least developed country
MAF	Ministry of Agriculture and Forestry (became MAFF, or Ministry of Agriculture, Forestry and Fisheries, in 1978)
MFA	Ministry of Foreign Affairs
MITI	Ministry of International Trade and Industry
MOF	Ministry of Finance
MSAC	most seriously affected country
ODA	Official Development Assistance
OECD	Organisation for Economic Cooperation and Development
OECF	Overseas Economic Cooperation Fund
OOF	Other Official Flows
OTCA	Overseas Technical Cooperation Agency

PARC   Policy Affairs Research Council (of the LDP)
UNCTAD   United Nations Conference on Trade and
                Development

PREFACE

This is a book about Japan's foreign aid and about how Japanese aid policy is made. Japan is no newcomer to the ranks of the international aid donor — aid has occupied, directly and indirectly, much of her international economic bureaucracy for nearly three decades. She is increasingly called upon to bear a greater share of the developed world's contribution to the developing nations, and her own foreign policy is now closely identified with the problems of aid policy. In late 1979 she is about to embark with China on a huge economic cooperation agreement that will have far-reaching effects on Japan's aid policy and on Western approaches to financing Chinese development.

This book describes the roots of Japan's aid policy, and shows that this side of her international economic policy is based largely on domestic conditions, structures and forces. To understand the pattern of Japanese aid as it has developed and as it stands today, it is important to appreciate the complexities of the Japanese decision-making process. It is hoped that this book can explain, both to the non-specialist reader and the aid practitioner, the patterns of Japanese aid policy-making.

The book owes much to many colleagues. Arthur Stockwin and Peter Drysdale (Australian National University) and Ide Yoshinori (University of Tokyo) all encouraged the work, while Fukui Haruhiro, John White and others gave valuable comments. My greatest thanks go to numerous officials in Japan for their consideration, graciousness and cooperation. Acknowledgement is also made to the publishers of *Australian Outlook* and *Pacific Affairs* for permission to draw on material first published in those journals. My wife Judy gave her strong support from the beginning.

The book was written entirely while I was working at the Australian National University. The views and interpretations are, of course, my own.

Alan Rix

Canberra

INTRODUCTION

During a speech in New York in May 1978, the Japanese Prime Minister, Fukuda Takeo, announced that his government would double its official development assistance within three years. That promise was reiterated at the summit meeting of heads of government of the advanced countries in Bonn in July 1978 and was later repeated by Mr Fukuda's successor, Ōhira Masayoshi, at the Fifth United Nations Conference on Trade and Development in Manila in May 1979. It has been taken up as the catchcry of those perceiving new purpose and direction in Japanese aid policy.

Mr Fukuda's pledge was the most forthright statement of aid policy in the years following the oil crisis, but it highlighted several contentious aspects of Japan's aid performance. Mr Fukuda implicitly acknowledged that both the quality and the quantity of Japan's aid were under question, and that improvements were urgently required. Certainly there were political reasons for the announcement: the Bonn summit and the mounting problems of trade surpluses and yen values demanded initiatives. Criticisms of his promise, however, were directed at the question of performance. *The Economist* commented tersely that it was only 'a tiny crumb for the Third World'.[1] The DAC, in its series of annual meetings in October 1978, was said to have 'derided' Japan's plan to double ODA in three years,[2] while in Japan it was recognised that the promise was probably of little value when disbursements were measured in dollars, a currency then falling rapidly against the yen.[3] Yet Japan's 1978 aid results revealed that in dollar terms her goal was already halfway to being reached.

Despite her many critics, Japan in the postwar years quickly became one of the world's largest donors. Foreign aid was always important to Japan's economic policy-makers: it was aid that helped Japan develop its trade and political links with postwar Asia; it was aid, given when Japan was herself striving to reach high growth in the 1950s and 1960s, that encouraged exports of heavy plant and equipment and helped to assure stable supplies of resources and raw materials. Aid was integral to Japan's available repertoire of foreign economic policies and for some was the only diplomatic weapon Japan could use in her relations with the developing countries.

The oil crisis of 1973-4 was a turning-point in this donor-centred approach to foreign aid. Continuing economic recession and only

11

gradual recovery complicated the pressures on aid officials and their budgets and compounded the structural problems of actually spending aid funds. Japan's dilemma as a resource-poor, trade-dependent economy in the volatile international circumstances of the late 1970s was one where the need to align aid more sensitively with Japan's overseas trade and monetary policies conflicted with the demands of the developing countries for economic assistance attuned to their own development needs. To many, Mr Fukuda's promise was more obviously designed to serve the first, rather than the second, objective.

The rapidly expanding programme drew mixed reactions over the years. Some aid donors and recipients did not acknowledge it as a true contribution to international aid. Technical assistance and other grant aid was relatively low, and the Japanese tended to overemphasise official export credits and private investment as part of the aid effort. The DAC, in its annual Aid Review, always found Japan's aid wanting by comparison with the efforts of other donors, but criticism from home and abroad did not have any lasting impact on the way Japanese policy was made. A somewhat imbalanced aid performance highlighted, in fact, some continuing problems of aid administration. Within a period of over twenty years Japan became one of the world's largest foreign aid donors, but failed to formulate an accepted set of policy guidelines to assist ministries in assessing aid requests. Many Japanese looked to foreign aid as a safeguard against apparent economic vulnerability, but basic agreement here was not carried over into the details of policy. Ministries involved with foreign economic policy justified aid in terms of their own vision of Japan's best interests.

It is to these issues that this book is addressed: how the Japanese aid effort was pushed and pulled by domestic conditions, how the contours of the aid programme were shaped by the decisions of the Tokyo-based bureaucracy, and on what premises and principles. It asks why the Japanese Government did not respond directly to the aid challenge, what influenced the allocation of Japan's aid, and why patterns persisted over many years. It argues that, because of serious organisational constraints, the character of the domestic aid administration decisively affected Japanese aid performance, for aid policies were inseparable from processes. Processes themselves were dominated by procedures but always through a changing policy environment where there were three important relationships:

1. between ideas in Japan about aid, and aid organisation and processes;

2. between the degree of priority enjoyed by aid within Japanese domestic politics, and the range and pattern of participation in policy-making;
3. between the interaction of officials, agencies and procedures on the one hand, and narrowed aid policy options for the Japanese Government on the other.

These relationships had a direct cumulative impact. Ideas about aid in Japan helped define its organisational presence. Conflicting perceptions of aid and its uses encouraged ministries, in the 1950s and early 1960s, to consolidate their influence in the aid administration, which all regarded as lying within their own jurisdiction. The result was diffuse structures which produced distinctive processes for different types of aid.

The low priority afforded aid within Japanese politics led to sporadic participation by ministers and a dependence on information and advice from outside government. Lack of sustained political interest in aid, except in the case of substantial bilateral relations, meant that staffing of aid institutions and administrative reform were given little attention, agencies were overloaded and high-level advisory bodies weakened. Structures replaced politics as the major influence on patterns of participation.

Predominantly bureaucratic participation impinged on Japanese policy choices. Budgeting imposed serious constraints on aid flows, and the generalist traditions of government personnel management limited their direction and content. An absence of policy guidelines led to coordination difficulties and an emphasis on detail and procedure. Government loans became the centre of policy and through the impact of development surveys and the politics of particular bilateral relationships directed aid flows into a self-reinforcing 'cycle' of aid to selected recipients.

A study of how Japan made aid policy leads into other questions. One is comparative: is Japan's case unique, how do other donor systems react to similar problems? Another relates to Japan specifically: to what extent were patterns detected for foreign aid common to other policy areas? Because the emphasis is on policy-*making*, however, it is therefore on the donor, and little is said of the recipient, of the effects of the Japanese programme or of the merits of particular types of aid as against others. To assess these issues carefully requires a sound understanding of a donor's motives and of the reasons for the way aid flows are structured. This book takes the first steps in that direction.

Foreign Aid and Japan

Argument about foreign aid in Japan involved problems which were, naturally enough, common to the aid debate generally. 'Foreign aid' is an elusive but controversial concept. The term can refer to certain economic phenomena (the 'explicit transfers of real resources to less developed countries on concessional terms'[4]) which admit of a variety of origins, commercial or official. It can be defined more narrowly as 'government-sponsored flows of resources made available on concessional terms to foreign governments',[5] where origin and point of receipt are more closely described. In both cases, foreign aid involves a movement of resources from one country to another in a way designed to assist the latter's development. It is generally accepted, furthermore, that this movement falls within the ambit of government policy to some extent.[6]

'Aid' is also used as a much broader category, where precision is lost and distinctions between kinds of resource flows are blurred. Here aid is lumped together with other foreign economic policies of governments and their subsidiary agencies. 'Foreign aid' as a defined exchange of concessional resources becomes equated with the more encompassing notion of 'economic cooperation'. In Japan, as we shall see, a precise concept of 'foreign aid' was not often distinguished from that of 'economic cooperation'. In this book it is accepted that the latter is a far more comprehensive category which includes non-explicit transfers of resources, such as those realised through trade and tariff policies. The term 'foreign aid' is used to refer to official development assistance, or ODA, as defined by the Development Assistance Committee (DAC) of the Organisation for Economic Cooperation and Development (OECD). This study is of Japanese official development assistance policy, not of economic cooperation policy, although the two overlap. Since the making of ODA policy affected other flows of assistance, we shall discuss these as they are relevant.

Perceptions of aid in Japan were regularly cited as the main influence on Japanese policies and the discussion of those ideas normally led into arguments about causes. Explanations of the nature of Japanese aid assumed that there was a strong consensus regarding aid in Japan, for reparations experience and the imperatives of domestic economic growth dominated Japanese approaches to the subject. The debate in the West about aid had little impact on Japanese policy-makers shaping their own priorities for aid in terms of the home economy. It is important, nevertheless, to realise the extent to which Japan's aid role was attributed to common perceptions of aid. The whole system, it was thought, was united. Hasegawa, for example, distilled a composite 'historical national evolutionist view', according to which aid 'is seen as an instrument of Japan's national policy to serve the *kokueki*, or national interest, of "secularized postwar Japan" '.[7] Edgar C. Harrell suggested that different groups influenced aid policies over the 1950s and 1960s because they agreed about the relationship between foreign aid and domestic economic growth. Priorities were agreed upon, elites made policy, and a 'divergence of views, when it surfaced, was generally about a specific project, country emphasis or the structure of aid administration'.[8] Many criticisms of Japanese aid policy (of which Halliday and McCormack's neo-imperialism analysis is one of the best known[9]) supported this 'consensus' view and attacked Japan's performance, usually in terms of a 'monolithic' approach to Japanese behaviour.

In practice, as the following chapters point out, policy as implemented was *not* an immutable whole; Japanese foreign aid was of many kinds, some more important than others, but her aid effort embraced no single 'aid philosophy', except appeals to the national welfare cast in predictable terms by each ministry. In this sense, aid policy was little more than a collection of disparate programmes supported by the self-serving slogans voiced by any donor.

Politics, Process and Policy-making

Questions of politics are important to most aid donors, not only Japan, although little has been done to identify the comparative aspects of donor policy-making.[10] Studies of donor administrations (of which there are a great many forms) support the view, stated by Viviani, that 'for the most part it is the way in which political responsibility for aid is exercised, together with the way bureaucratic control is located that decisively shapes an aid program'.[11] Political and organisational aspects are assessed by Judith Tendler in her study of the impact of organisation on the tasks of the United States Agency for International Development (AID). She concluded that political pressures impinged on the organisation as criticism and as incursions by other public agencies into the territory of AID. 'Criticism and incursion', she wrote, 'affected the agency's performance by changing the AID technician's concept of what he wanted to do.'[12] In Japan, much of the politics of foreign aid took place within the bureaucracy and low levels of political support for aid reinforced this tendency.

The link between politics and participation was the subject of much of the debate about Japanese policy-making, although the relevance of the relationship to the bureaucratic process was not usually discussed. Japanese scholars once took a static institutional approach in their studies of the national bureaucracy, but more recent writers stressed process rather than structure and, in doing so, came up with some significant conclusions about Japanese policy-making as a whole.[13] Fukui showed in his survey of the literature on policy-making in Japan that the emphasis of earlier writings by Japanese scholars on the power of the LDP or, alternatively, of the ruling elite of party, business and bureaucracy had given way to 'a picture of Japanese policy-making which is characterized above all by fluidity, complexity, and variability, rather than by the regularity, stability and constancy which the power-elite perspective projects'.[14] In the aid policy area, as this book demonstrates, policy was not simply a reflection of prevailing political power structures or of social groupings. Rather, the organisational process of routine bureaucratic decision-making proved decisive.

John White, in his study of the politics of foreign aid, concluded that 'the makings of an aid policy lie in the hands of those who actually administer it'.[15] How aid management procedures channelled bureaucratic politics and affected actual policy content is central to this book, which in a sense takes up where John White left off. The aid administration and aid administrators were variables in the aid relationship which cannot be ignored. Another such factor was the way in which implementing aid decisions limited future policy alternatives. This problem is often discussed in relation to the developmental impact of aid on recipients, but the effects of feedback on the *donor* policy process are taken up in later chapters.

However important the organisational dimension, outcomes are also affected by individuals. Policy is derived from the actions of individuals in certain institutional patterns, and the organisational environment in Japan and its constraints on individual actions and interactions are dominant themes in this book. Standard operating procedures and incrementalism are as much a part of Japanese as of Western bureaucracies, although the emphasis here on organisation offsets what might appear to be a persistent 'bureaucratic politics' theme: it is hard to allow as much scope for individual influence as Allison's original bureaucratic politics paradigm suggests.[16] Patterns of participation are the outward evidence of policy-making, but in this book we are not so much interested in ideal types of policy-making processes as in the dynamics of policy and the roots of that dynamism.

The book is divided into three parts. The first two chapters look at ideas about aid in Japan and the development of aid organisation. Chapter 1 traces the growth of Japanese aid adminstration and the changing place of aid within national priorities, while Chapter 2 details one aspect of the relationship between aid structures and policy, the creation of the Japan International Cooperation Agency.

Part II is about aid politics in Japan, the interaction between administrative behaviour and traditions, procedures and the imperatives of annual budgeting. Chapter 3 outlines the various elements of the domestic aid system, Chapter 4 describes decision-making for different types of aid, and Chapter 5 analyses the influence of aid budgeting on policy. Part III looks outward to Japan's aid relations as a whole. Chapter 6 first shows how domestic procedures react to the recipient end of the aid relationship, and how the role of private enterprise is essential in maintaining the momentum of the aid process. Chapter 7 discusses the patterns of Japanese aid policy, how they are maintained by persisting emphasis on certain bilateral relationships. There are important undercurrents which directed policy also, and as well as being part of the diplomacy of aid, they were related to administrative problems. Chapter 8 analyses the role of information in aid policy-making, the central function of the executive agencies in prolonging the self-reinforcing 'cycle' of aid along well-worn paths, and finally the bases of growth and change in the aid programme.

Notes

1. *The Economist*, 6 May 1978, p. 95.
2. *Japan Economic Journal*, 17 October 1978.
3. *Yomiuri shimbun*, 1 July 1978. The US dollar fell dramatically over a period of several months in 1978, from a value of 264.5 yen in September 1977 to 190 yen in August 1978. All figures quoted will be in US dollars unless otherwise indicated.
4. Jagdish Bhagwati and Richard S. Eckaus (eds.), *Foreign Aid* (Harmondsworth, Penguin, 1970), p. 7. Raymond F. Mikesell in his *The Economics of Foreign Aid* (London, Weidenfeld and Nicolson, 1968), presents a full analysis of foreign aid in economic terms.
5. David Wall, *The Charity of Nations: The Political Economy of Foreign Aid* (London, Macmillan, 1973), p. 3.
6. It is the opinion of this writer that foreign aid (except military assistance) can provide an important complement to the resources available for the development of the developing countries. This is, however, provided that donor policies take full account of recipient defined needs and priorities and that the recipient is as free as possible to use those resources in accordance with its own goals.

7. Hasegawa Sukehiro, *Japanese Foreign Aid: Policy and Practice* (New York, Praeger, 1975), p. 7.

8. Edgar C. Harrell, 'Japan's Postwar Aid Policies', unpublished Ph D dissertation, Columbia University, 1973, p. 269.

9. Jon Halliday and Gavan McCormack, *Japanese Imperialism Today: Co-prosperity in Greater East Asia* (Harmondsworth, Penguin, 1973).

10. 'Policy' is used in this thesis as Heclo defined it, to 'designate a course of action or inaction pursued under the authority of government' (Hugh Heclo, *Modern Social Politics in Britain and Sweden: From Relief to Income Maintenance* (New Haven, Yale University Press, 1974), p. 4). The term includes both intentional 'policy outputs' and unintended 'policy outcomes'.

11. Nancy Viviani, 'Problems of Aid Administration and Policy Formulation Among Western Countries', unpublished paper, Canberra, Australian National University, 1977, p. 1.

12. Judith Tendler, *Inside Foreign Aid* (London, The Johns Hopkins University Press, 1975), p. 47.

13. For example, Kawanaka Nikō, 'Nihon ni okeru seisaku kettei no seiji katei' ('The political process of policy-making in Japan'), Ide Yoshinori, 'Gyōsei kokka ni okeru "kan" no shihai' ('The role of the official in the administrative state') and Kojima Akira, 'Gendai yosan seiji shiron: wagakuni ni okeru yosan katei no hōhōronteki kōsatsu' ('An essay on contemporary budget politics: a methodological study of the Japanese budget process'), all in Taniuchi Makoto *et al.*, *Gendai gyōsei to kanryōsei* (*Contemporary administration and the bureaucracy* (Tokyo, Tōkyō daigaku shuppankai, 1974).

14. Fukui Haruhiro, 'Studies in Policymaking: A Review of the Literature' in T.J. Pempel (ed.), *Policymaking in Contemporary Japan* (Ithaca, Cornell University Press, 1977, p. 47).

15. John White, *The Politics of Foreign Aid* (London, Bodley Head, 1974), p. 3.

16. Graham T. Allison and Morton H. Halperin, 'Bureaucratic Politics: A Paradigm and Some Policy Implications', *World Politics*, vol. 24, (Spring 1972), pp. 40-79; and also Graham T. Allison, *Essence of Decision: Explaining the Cuban Missile Crisis* (Boston, Little, Brown and Company, 1971).

Part I:
Aid Ideas and Aid Structures

1 FOREIGN AID AND THE MINISTRIES

Japan's relations with the developing nations had achieved by 1979 an importance in foreign economic policy not seen in three decades of vigorous trade and economic ties with the developing world. The Japanese Prime Minister, Ōhira Masayoshi, addressed the Fifth United Nations Conference on Trade and Development in Manila in May and received a warm response for his promises of more and better aid, support for the Common Fund for international commodity stabilisation, improved access for developing country exports, and renewed efforts to develop manpower policies for the Third World. This was in marked contrast to the strongly critical reaction of the poorer nations to Japan's lacklustre performance at the first UNCTAD fifteen years earlier. By June 1979 official DAC figures showed that Japan was well on the way to achieving her goal of doubling Official Development Assistance (admittedly in dollar terms) between 1977 and 1980, and it appeared that Japan's foreign economic policy-makers were committed to a larger, more positive contribution to solving the North-South issue.

It had not always been so. Japanese foreign aid policy was always strongly influenced by the thinking of government officials, a group of men not easily united in their views on aid and economic cooperation. Since the 1950s, ministries and agencies found themselves in conflict about many of the fundamentals of aid policy — the best form of aid, the direction, size, methods and terms of aid flows, the precise objectives of aid and the nature of relationships between aid and other domestic and foreign policies. In their arguments about policy, ministries often appealed to similar ideals (the duty and responsibility to provide aid, fairness and mutuality in aid relations) but the end result, rather than the motivation, was always decisive in domestic argument on aid questions. These ends tended, however, to be viewed in the short term and the shifting bureaucratic battle over aid policies and the administrative control of aid were founded neither on lasting consensus on policy objectives nor on agreement about what Japanese foreign aid could in practice achieve.

This first part of the book takes up the argument that underlying differences in the way ministries perceived aid and its uses led to a hardening of approaches and of administrative structures by the mid-1960s which persisted well into the 1970s. This chapter traces official attitudes to foreign aid and, briefly, the development of the aid bureaucracy, while Chapter 2 examines a particular example of the conflict between ministry policies for aid, the establishment of the Japan International Cooperation Agency.

This distillation of official attitudes took place over five distinct but overlapping periods, beginning with the years of reparations and minor technical assistance until the first yen loan was made to India in 1958. Then followed the period until the early 1960s when the Development Assistance Committee was set up. Horizons broadened throughout the 1960s until the Second United Nations Conference on Trade and Development (UNCTAD) in 1968, as Japan became accepted as one of the world's economic powers and took on more of the responsibility attendant on that status. The height of the Japanese aid effort came between the late 1960s and the oil crisis of 1973, but the ensuing economic recession lasting into the latter half of the 1970s produced some rethinking of Japanese policies and the first steps, such as the commitment to double aid in three years, to regain the momentum of earlier years and to reassess the relationship between aid and other national policies.

Economic Cooperation and Trade

The first tentative steps in Japan's foreign aid policy were very modest by comparison with the record of the 1960s and the large disbursements of the late 1970s.[1] Emerging from its postwar diplomatic cocoon during the 1950s, Japan initially gave foreign aid in the form of technical assistance, and later entered a number of reparations agreements, both to fulfil outstanding obligations and to serve the export potential of her renascent industries. The first official loan was made in 1958.

Technical aid derived from Japan's participation in international technical cooperation schemes, such as the United National Expanded Programme of Technical Assistance and the Colombo Plan. Japan concluded reparations agreements with Burma in 1955, the Philippines in 1956 and Indonesia in 1958, while quasi-reparations (grants in lieu of formal reparations commitments) were arranged with Laos and Cambodia in 1959.[2] Nor was Japan's own economic well-being neglected: the establishment of the Export-Import Bank of Japan (Eximbank) in 1952 signalled the beginning of full-scale government assistance to exports and encouragement to private investment overseas. Total official net flows rose from an average of $10 million between 1950 and 1955 to \$285 million in 1958.[3]

In this period to 1958, the concept of 'economic cooperation' was used by government officials and businessmen alike to describe Japanese attempts to further economic relations with developing nations, especially those of Asia. The distinction between this notion and 'aid' was not made, and the two terms were more often confused than clarified. Different ministries separately defined economic cooperation but never explicitly identified an aid component. The Ministry of International Trade and Industry (MITI), the Economic Planning Agency (EPA) and the Ministry of Foreign Affairs (MFA) each drew up its own set of priorities into which economic cooperation was fitted, the common factor being that Japan's domestic economic prosperity, was placed foremost. In their separate approaches to these policy problems, emphases overlapped but distinctions remained.

The EPA in these early days relied on broad definitions and made little mention of 'aid'. The first postwar economic plan of 1955, prepared by the EPA and aimed at establishing a national programme for Japan's postwar economic growth, included economic diplomacy and trade promotion in Asia as two of its important goals. The second plan of 1957 took these further, by expressing in more detail the long-term perspective of Japan's economic planning and making explicit the need to coordinate domestic and international economic strategies. Measures to redirect Japan's industrial structure from light industry to heavy and chemical industries depended on the promotion of trade to the developing markets of Southeast Asia by means of international economic policies, especially economic cooperation. The plan defined economic cooperation in sweeping terms to include development planning, capital exports, the extension of credits, overseas investments, the procurement of resources and long-term import policies for food and raw materials.

In the first edition in 1958 of MITI's *Economic Cooperation: Present Situation and Problems*,[4] the only full annual report on Japan's aid policies until 1978, economic cooperation was blatantly linked, although without the long-term perspective of the EPA, to the goal of Japanese trade promotion. The MITI report presented the view, expressed early in the 1950s by those in business and government anxious to press ahead with external economic relations,[5] that Japan's main economic task was the promotion of trade to secure the resources necessary for Japan's industrial growth, and to develop markets for the products of Japan's industry. Economic cooperation became, in the words of the MITI paper, 'the new axis of postwar trade policy', for the developing nations were valued as markets or as potential markets in which demand could easily be stimulated.

While this object was clear-cut, it did not help to clarify terminology. The 1958 report stated that the words *keizai kyōryoku* (economic cooperation) could refer not only to relations between developed and developing nations, but also to those between developed nations themselves. MITI focused on the former and claimed that economic cooperation involved a very wide (and ill-defined) area of both government and private participation in the development of the 'underdeveloped countries'. This cooperation could, according to the report, be either bilateral or multilateral, and included capital, technical and trade flows. No reference to aid (*enjo*) was made.

The first foreign policy review published by the MFA in 1957 planted economic cooperation firmly within the scope of Japan's national policies and international interests but, more so than MITI's report, considered the strategic and security implications.[6] Like MITI and the EPA, the MFA considered economic cooperation to be necessary for Japan's own growth, and it openly acknowledged the significance of Asia in this respect: 'Our own development is not assisted by an Asia without peace, progress and prosperity.' The Southeast Asian nations were regarded as increasingly important sources of food and raw materials for Japan, as well as markets for Japanese exports.

Specifying where economic cooperation would be best directed still did not explain what the concept meant. In this first review, the MFA made a clumsy attempt to define economic cooperation. At one point, it was said to comprise reparations and foreign aid, since countries not claiming war damages, it was suggested, were entitled to receive foreign aid. The 1957 review stated later that economic cooperation included reparations, technical cooperation, private cooperation and government assistance to private business, such as that undertaken by the Eximbank. What precisely constituted aid, however, was not explained. Among these elements of economic cooperation, the MFA gave priority to the pursuit of private cooperation and technical assistance, because it judged, with uncharacteristic frankness, that only limited resources were available for government capital assistance. This candour suggested that the MFA even in 1957 regarded aid as government flows of capital excluding reparations, which were included in a more extensive economic cooperation effort by both government and business.[7]

Differences in approach had immediate, practical significance.

Acting largely independently, but out of a shared awareness of the externally oriented future of the Japanese economy, four ministries had by 1955 placed themselves squarely in those areas of the national administration destined to become the crux of economic cooperation policy from the late 1950s. The MFA, through its economic affairs and reparations divisions, had the main part in managing reparations agreements and Japan's relations with multilateral agencies and the wider developing world. The MOF was responsible for international monetary policy (and, of course, budgeting), MITI's expanding trade administration was the controlling authority in exports and trade with the developing countries, and the EPA, set up in 1955, undertook overall economic planning.

By the end of the 1950s Japan was deeply involved in economic cooperation. Private investment and technical assistance, both official and private, formed the core of policy, with Southeast Asia as Japan's main target. Government-to-government reparations agreements strengthened the geographical bias. The economic and diplomatic interests of several ministries were directed towards cultivating close and potentially large markets, for it was hoped that regional economic development would ensure the economic, social and political stability of Asia, which many Japanese officials considered essential to peace in the region. In foreign policy, Prime Minister Kishi Nobusuke's active diplomacy and his initiatives towards improved Japanese relations with Southeast Asia demonstrated a keen appreciation of how economic cooperation would affect Japan's future. Economic cooperation became an integral part of Japan's long-term policies, and government initiatives were fully supported by the economic diplomacy of Japanese business.[8]

Yen Loans: Extending Horizons

In the late 1950s Japan's economic cooperation expanded significantly but official attitudes to the issue diverged further. The Japanese Government granted its first yen loan in 1958 as part of contributions by the World Bank Consortium for India.[9] That nation was then regarded in many countries as the cornerstone of stability in Asia and the consortium of which Japan was a member fully endorsed this view.[10] Other early government loans were made to Paraguay, South Vietnam, Pakistan and Brazil but, while private investment and export credits rose in the period 1958-61 from $33 million to $160 million, total net official flows in fact fell from $285 million to $221 million, due largely to a fall in grants.[11] The period was remarkable not so much for the changes in the direction of economic cooperation but for the hardening of institutional arrangements and a greater awareness of economic cooperation and aid arguments in Japan. New objectives and new modes of cooperation replaced reparations as the central feature of Japanese policy. The demands of trade promotion and a long-term Asian policy remained strong, but so did Japanese attempts to be accepted in Europe and America as a responsible trading partner, and to foster an image which could enhance Japan's claims to membership of the principal international economic organisations.

MITI showed some appreciation of the new direction when it referred in its 1960 review to economic cooperation as 'the mission of the world's industrial nations'. In practical terms, however, this moral goal was to be achieved not by highly concessional assistance but by increasing direct loans and investment overseas to secure vital raw materials from resource-rich developing nations through policies of 'development import'. MITI's recommendations for increasing the quantity and quality of economic cooperation were designed primarily to further not the development of the recipient nations but Japan's own trade and economic policy ends. The 1961 report enlarged on this by stating that Japan undertook economic cooperation to develop her own industry. She did not do so out of support for Cold War political objectives or support for development as part of decolonisation policies.

This oblique and mildly critical reference to the aid policies of Western donors was the first overt comparison by the Japanese Government of her own and Western efforts. It marked also the first defence of Japanese economic cooperation in terms of the aid debate then fashionable in the United States and Europe, but MITI seemed to have little compunction in dismissing the development claims of the Third World. 'Aid' as a separate endeavour was ignored.

The MFA, in contrast, appeared more aware of Japan's unique position as an aid donor and was able to present a fairly sophisticated case for Japan. The terms 'economic cooperation' and 'aid' were still confused, on occasion being used alternatively in respect of technical assistance, but by 1961 the distinctions were more pronounced in ministry statements. It was suggested then that official flows represented the 'aid' component of a complex pattern of official and private transactions.

The MFA put forward three related arguments in support of Japan's economic cooperation. As a first step, it justified the persistent bias towards Asia by citing interdependence between the Japanese and the Southeast Asian economies and the need to ensure their stability and prosperity. It then took the argument beyond Southeast Asia and introduced the North-South debate and US-Soviet relations, by saying that aid (*enjo*) should be increased because some nations were politically unstable. This in turn led the MFA to maintain that the moral basis of Japan's economic cooperation, which was first provided by commitment to reparations, was still firm. It saw a 'natural role' for Japan as Asia's largest economy in world attempts to alleviate the development problems of the Third World.

It was doubtful, however, whether the MFA took all these rationalisations seriously, for the minstry was obviously aware of the constraints of Japan's largesse in those early years, while still being keen to pursue the benefits it foresaw in an enlarged programme. The 1959 foreign policy annual continued to defend the small size of Japan's economic cooperation in more parochial ways by claiming that Japan's economic capacity was limited. It offered only increased technical assistance, mainly as a means of lightening the impact on Japan's meagre finances. The 'natural' responsibility which Japan recognised was only natural, it was admitted two years later, 'because we depend on them [the developing nations] for 45 per cent of our trade'.

This uneasy blend of altruism and 'good housekeeping' in economic cooperation was reinforced by the Income Doubling Plan of 1960. It was this document which attempted to set the pattern for the next decade of economic activity and which linked domestic economic policy and economic cooperation to achieve the best use of domestic resources. While it placed foreign economic policy at the centre of Japanese policy, the plan relegated economic cooperation as such to a subservient position. It diverted Japanese attention from the widening aid debate, the imperatives of the incipient United Nations Development Decade and hopes for a positive contribution to the Development Assistance Group (DAG) and the DAC. These aspirations were overshadowed by an introspective policy emphasis, an international policy based on unyielding domestic interests and an aid effort divorced from concern for the welfare of the developing nations. The plan appraised Japanese aid in terms of its prospects for quantitative growth rather than qualitative achievement, a standard which was never displaced and which found echo still in the 1978 promise by the Japanese Government to double its ODA in three years. In short, Japan entered the 1960s determined to undertake economic cooperation, but only to the extent of her national capabilities and primarily for trade development, domestic economic prosperity and broad international political objectives.

Japan and the DAC

In 1961 Japan joined, as a founder member, the Development Assistance Committee of the OECD and thus made a place for herself among the aid-giving nations of the world. This move showed that other nations, especially the United States, encouraged Japan in aiding Asian nations, and that Japan herself recognised this task. It revealed also Japan's increasing determination to be accepted as a responsible member of the club of advanced nations, despite criticism of Japanese development assistance policy from other donors.

Japan's membership of the DAC has been referred to by one writer with experience in DAC affairs as 'an apparent historical anomaly'.[12] This judgement would seem to have some truth in it. In 1960 the Development Assistance Group was established in the Organisation for European Economic Cooperation (OEEC). It was an *ad hoc* meeting, the result of increasing American pressure on its allies to join in a coordinated international effort on aid. In 1961, when the OEEC became the OECD, the DAG was reconstituted as the Development Assistance Committee. Nine founder members of OECD (Belgium, Canada, France, West Germany, Italy, Netherlands, Portugal, UK, USA) joined, as did Japan, although she was not a member of the OEEC and did not enter the OECD until 1964.

Japan joined originally, it seemed, in an attempt to give herself 'a foot-hold in the group of the more powerful states and greater influence in both world and regional affairs where it had both commercial and political interests'.[13] She achieved some of these aims, but at the same time became subject to greater scrutiny of her aid programmes. The *Nihon keizai shimbun* recognised the possible restrictions and obligations involved in membership, but acknowledged that joining 'would help bring [Japan] out of her isolation'[14] and would be important in promoting Japan's membership of the OECD.[15] Nevertheless, Japan remained cautious and was careful not to commit herself on aid policy. The government representative, Shima Shigenobu, went to Washington on 8 March 1960 with 'no intention of making clear Japan's detailed plans for foreign aid'. His purpose, as reported in the *Nikkei* of the following day, was instead to 'show Japan's face' and 'press for participation in the new aid organisation'. The Director of the MFA's Economic Cooperation Division, Sawaki Masao, made it clear that Japan joined mainly from a desire to avoid long-term disadvantage by getting a foot inside the OECD door. European nations were, in his opinion, still opposed to Japan's entry to that body, but membership of the DAG could give her a valuable link to the parent organisation. America was seeking the solution of aid problems 'at a world level' and this assisted Japan, especially since she would be the only Asian nation to join the DAG. Sawaki also admitted quite frankly that she had asked the US Government if she might join only after satisfying herself that the DAG was not a body which bound its members to agreed aid commitments. He believed that the greatest benefit arising from Japan's entry would be an understanding of the policies of the other donor nations, which Japan had been trying unsuccessfully to gain for some time.[16]

Within the group, the United States concentrated on persuading West Germany and Japan of their special obligations as former recipients of Allied aid and as nations which spent little on defence. At the DAG meeting in London in March 1961 the US asked for a group target for aid to be set at 1 per cent of GNP. The Japanese MFA stressed back home that Japan was not necessarily required to produce this amount, although she did recognise the need for expansion of present aid budgets. A ministry statement in the *Nikkei* of 28 March envisaged Japan reaching such a target only by retaining private investment as the core of any policy.

MITI, in a later assessment of the DAC, pointed to the American policy of 'spreading the burden' as a strong influence which emerged in the early meetings of the DAG. Although Japan claimed independence of this policy, MITI, like the MFA in earlier years, contended that the government joined out of serious motives: a desire to avoid isolation from other advanced nations, to be close to those countries working together in the OEEC and to bring the benefits of DAG membership to its own economic cooperation and trade policy.[17]

Japan at this early stage seemed concerned that she would not be allowed to determine her own level of aid. The Fifth DAG Meeting was held in Tokyo in July 1961. In a speech to the group, the Prime Minister, Ikeda Hayato, declared that Japan would cooperate to the limit of her ability, but gave no firm promises. The July meeting agreed that Japan should join the DAC, but it did not alter its reservations about her participating in the OECD. Japan therefore set up a DAC liaison office in her Paris Embassy in October 1961 and announced her intention to cooperate as fully as possible with the DAC policy of greater aid contributions, although one newspaper suggested rather cynically that this positive expression of support was prompted by the need to encourage machinery exports.[18] As a member of the DAC, Japan was given the right to attend and speak at OECD Directors' Meetings when DAC items were under discussion. Membership of the OECD came three years later, an event related partly to the desire of the OECD 'to induce Japan to accept a larger responsibility for aid' but also to the much more general issues of trade liberalisation and the place of Japan in international trade.[19] In a sense, the conflict between Japan's initial reluctance to be bound by DAC commitments to increase aid and her enthusiasm of OECD participation was resolved by having to accept that membership brought the risk of criticism from fellow members.

Entry to the DAG and then DAC affected Japan's aid effort in several ways. Although it is likely that not even MFA officials fully realised the long-term significance of Japan's membership, Japan was able to evaluate other donors' programmes and assess her own in the light of this knowledge. In addition to the satisfaction from this, however, she was forced into the mainstream of the aid debate centred on the DAC and was made aware, sometimes embarrassingly, of her responsibilities as a donor. This awareness encouraged a sense of competition with other members and thereafter Japanese official reports on economic cooperation were careful to note Japan's usually improving relative position in terms of quantity of aid given. Quality featured less prominently.

Membership of the DAC inadvertently helped the Japanese to clarify the usage of the terms 'economic cooperation' and 'aid', although this took some time and did not erase all differences in the way ministries adopted the words.[20] Throughout the 1960s, the Japanese usually distinguished 'economic cooperation' and 'aid' as concepts, but did not carry the distinction very far. The two were often used interchangeably. MITI economic cooperation reports consistently argued that 'economic cooperation' in its broadest sense consisted of three kinds of cooperation — capital, technical and trade. Japan's annual economic cooperation was determined by adding the value of all capital, most technical and a fraction of trade cooperation, to give the 'flow of financial resources' as defined by the DAC.[21] MITI insisted that while the DAC category was termed 'aid', it was only 'economic cooperation' in the narrow sense. True 'economic cooperation', claimed MITI in its 1970 report, took account also of trade policies towards the developing nations.

The general thrust of thinking of the 1950s continued into the 1960s. Japanese officials came to use the word 'aid' as a synonym for one type of 'economic cooperation', but carefully laid any blame for misuse at the feet of the DAC. They treated 'aid' (*enjo*) as that part of Japan's economic policies towards developing countries defined by the DAC as 'development assistance', government and private flows included, and it was not until late in the decade that a further line was drawn separating official and non-official resources. 'Economic cooperation' in the wide sense was retained to describe the full sweep of Japan's foreign economic policies. In this way, terminology was isolated from the debate about motivations.

The 1960s: An Expanding Japanese Role

The foreign aid given by Japan in the 1960s was quite different in scope and quantity from that provided in the 1950s. Over the period 1960 to 1968 the total net flow of official and private resources from Japan to the less developed countries increased from $246 million to $1,029.8 million (see Table 1.1) and in 1968 Japan became the fourth largest donor in the DAC. Official flows did not, however, rise as quickly, increasing from only $145 million to $356.2 million. In relation to GNP, which grew rapidly in the 1960s, ODA rose only from 0.24 per cent in 1960 to 0.25 per cent in 1968, although total flows increased as a percentage of GNP from 0.57 per cent to 0.73 per cent.

The size of total flows continued to depend heavily on private assistance, especially on export credits and direct investment. ODA fell as a proportion of the total, from 58 per cent in 1961 to only 34.5 per cent in 1968. From 1968, the category of Other Official Flows (OOF) was introduced by the DAC into the statistical presentation of aid flows and, when earlier figures were reconverted, Japan stood out as the member with the highest OOF portion of total flows. Furthermore, Japan's ODA as a percentage of GNP ranked very poorly, being the seventh lowest of all members in 1960 and the fifth lowest in 1968.

Within ODA, however, direct loans increased from $48 million in 1960 to $191.3 million in 1968, by far the largest rise of any ODA category. Bilateral grants and grant like flows (including reparations) rose from $67 million to $117 million and contributions to multilateral organisations from $30 million to $48 million. Technical assistance increased nearly six-fold from $2.4 million in 1961 to $13.7 million in 1968. Technical assistance and grants as a whole remained small even

Table 1.1: The Net Flow of Financial Resources from Japan to Developing Countries and Multilateral Agencies, 1960-78 ($ million)

	1960	1961	1962	1963	1964	1965	1966	1967	1968	1969	1970	1971	1972	1973	1974	1975	1976	1977	1978
Total Official & Private, net	246	381.4	285.8	278.4	303.8	485.5	538.8	855.3	1029.8	1263.1	1823.9	2140.5	2725.4	5844.2	2962.3	2879.6	4002.6	5534.9	10703.5
1. ODA, net	145	221.4	167.8	140.4	115.7	243.7	285.3	390.6	356.2	435.6	458.0	510.7	611.1	1011.0	1126.2	1147.7	1104.9	1424.4	2215.4
(a) Bilateral	115	210.0	160.6	128.2	106.2	226.3	234.7	345.9	308.3	339.7	371.5	432.0	477.8	765.2	880.4	850.4	753.0	899.3	1531.0
(i) Grants,	67	67.8	74.6	76.7	68.7	82.2	104.7	138.4	117.0	123.4	121.2	125.4	170.6	220.1	198.6	201.7	184.9	236.7	383.4
incl. Technical	3	2.4	3.4	4.5	5.8	6.0	7.6	11.0	13.7	19.0	21.6	27.7	35.6	57.2	63.5	87.2	108.1	147.8	221.2
(ii) Loans	48	142.2	86.0	51.5	37.5	144.1	130.0	207.5	191.3	216.2	250.3	306.6	307.2	545.1	681.8	648.7	568.1	662.6	1147.6
(b) Multilateral	30	11.4	7.2	12.2	9.5	17.4	50.7	44.7	47.9	95.9	86.5	78.7	133.3	245.8	245.8	297.3	352.0	525.2	684.4
2. OOF, net*			(82.9)	(35.9)	(94.9)	(109.7)	(182.7)	(198.9)	322.1	375.8	693.6	651.1	856.4	1178.9	788.9	1369.4	1333.4	1622.6	2152.6
3. Private, at market terms	101	160.0	118.0	138.0	188.1	241.8	253.5	464.7	351.5	541.7	669.4	975.6	1252.3	3647.5	1038.5	352.4	1548.1	2487.9	6335.5
Total/GNP (%)	0.57	0.71	0.49	0.40	0.36	0.55	0.62	0.67	0.73	0.76	0.93	0.92	0.95	1.44	0.65	0.59	0.72	0.80	1.09
DAC Average (%)	0.89	0.95	0.80	0.76	0.79	0.77	0.71	0.74	0.80	0.75	0.74	0.74	0.82	0.79	0.81	1.05	0.97	1.05	1.00
ODA/GNP (%)	0.24	0.20	0.15	0.20	0.15	0.28	0.28	0.32	0.25	0.26	0.23	0.23	0.21	0.25	0.25	0.23	0.20	0.21	0.23
DAC Average (%)	0.52	0.53	0.52	0.51	0.49	0.44	0.41	0.42	0.38	0.36	0.34	0.35	0.24	0.30	0.33	0.35	0.33	0.31	0.32

*The category of Other Official Flows (OOF) was first introduced in 1968. The bracketed figures before that date represent calculations by DAC of the value of OOF in those years. They comprise mainly official export credits.

Sources: 1960-2   OECD, *Flow of Financial Resources to less-developed Countries* (Paris, 1964).
         1963-6   OECD, *Development Assistance Efforts and Policies of the Members of the Development Assistance Committee* (or DAC, *Review*), 1966 and 1967.
         1967-70   DAC, *Review*, 1968, 1970 and 1971.
         1971-6   DAC, *Review*, 1973, 1976 and 1977.
         Ratios 1960-70   DAC, *Review*, 1971.
              1971-6   DAC, *Review*, 1973, 1976 and 1977.
         1977-8   Ministry of International Trade and Industry, *Flow of Resources from Japan to Developing Countries during 1978*, 31 July 1979.

in the late 1970s but, as a percentage of ODA in 1968, Japan's technical assistance was only 3.8 per cent compared to the DAC average of 23.3 per cent. On the other hand, multilateral contributions as a percentage of ODA in 1968 were higher than the DAC average, 13.4 per cent as compared with 10.8 per cent. Japan's aid-giving performance remained poor when compared with other DAC members, her relative position in the grant element table being eleventh (out of 15) in 1965 and the same (out of 16) in 1969. One feature of Japanese aid in the 1960s which continued into the 1970s was the low concessionality of Japanese official lending, and Japanese terms for official bilateral loans were always harder than DAC averages.

The pattern of geographical distribution of Japanese aid was established in the 1960s. While investments in the 1950s showed a marked emphasis on Central and South America and the Middle East,[22] official development assistance later became heavily concentrated in Asia. In 1963, 56 per cent of total flows was directed to Asia, and 98.7 per cent of ODA. In 1969, 73.8 per cent of total flows went to Asia and likewise 100 per cent of ODA, of which 48.5 per cent was to Southeast Asia (see Table 1.2). Japan's pursuit of an Asian foreign policy in the 1960s was amply supported by flows such as these.

These figures reflected the objectives of Japanese economic cooperation and aid policy in the 1960s. They showed also that the agreement between domestic ministries on the broad outlines of aid policy in the 1950s was weakened by the expansion of the aid programme in the following decade, particularly in the face of mounting criticism by both recipients and other donors. Heavier regional responsibilities, such as those which arose from Japan's participation in the Asian Development Bank (ADB), the Inter-Governmental Group on Indonesia (IGGI) and the Ministerial Conference on the Economic Development of Southeast Asia (MEDSEA), all initiated in 1966, exacerbated differences between the Ministry of Finance (MOF) and the MFA in particular, mainly because of the MOF's reluctance to sanction growth in the aid budget of the order that the MFA argued the new activities required.

In the 1950s all the ministries concerned with economic cooperation in the wider sense were aware of its potential benefits to Japan. They appreciated how economic cooperation contributed to Japan's own prosperity and trade promotion and, prompted initially by Japan's reparations obligations, all had established sections to administer different parts of the programme. International commitment, however, was soon replaced by the vested interest of ministries in organisational growth and their own policy responsibilities. Thus the MFA tended to

Table 1.2: Geographical Distribution of Net Flows of Financial Resources from Japan to Developing Countries (Percentage)

		Asia	of which Southeast Asia	Near & Middle East	Africa	Central & South America	Europe	Other
	1963	98.7	93.0	0.3	0.3	0.4	–	0.3
	1965	98.1	90.1	0.2	0.6	0.6	0.1	0.2
	1967	97.6	71.5	0.2	0.8	0.7	0.1	0.6
ODA	1969	100.0	48.5	0.8	1.2	-3.9	0.8	0.2
	1971	98.4	51.2	0.9	3.0	-2.6	-0.2	0.5
	1973	88.1	53.8	0.1	2.6	4.6	3.0	1.3
	1975	75.0	50.1	10.6	6.9	5.6	0.0	1.9
	1977	60.0	na	12.4	17.1	8.8	0.5	1.2
	1963	33.5	22.8	14.1	29.1	13.5	8.7	1.0
	1965	32.3	26.4	3.2	37.6	17.0	9.9	0.1
	1967	33.8	5.6	14.8	42.7	9.7	-1.0	–
OOF &	1969	62.5	21.5	13.1	6.4	11.2	6.8	–
Private	1971	52.3	11.8	7.3	10.5	23.2	4.2	2.6
	1973	30.4	15.8	3.2	8.5	53.4	3.0	1.4
	1975	53.0	38.5	14.0	6.7	24.8	1.1	0.4
	1977	17.8	na	14.6	25.0	30.0	12.4	0.2
	1963	56.0	34.4	8.0	16.4	10.3	4.9	4.4
	1965	53.4	27.8	1.9	22.4	13.4	5.9	2.9
	1967	58.5	25.2	8.1	23.3	5.2	-0.4	5.3
Total	1969	73.8	29.5	9.5	4.9	6.7	5.0	0.1
	1971	64.1	21.8	5.6	8.6	16.6	3.0	2.1
	1973	39.1	21.5	2.7	7.6	46.1	3.0	0.2
	1975	60.3	42.4	12.9	6.8	18.4	0.7	0.9
	1977	28.0	na	13.8	22.9	25.2	9.7	0.4

ODA =   Official Development Assistance.
OOF =   Other Official Flows.
na   =   not available.
Sources:
1963-7:   Gaimushō keizai kyōryokukyoku, *Wagakuni no shikin no nagare no chiikibetsu bunpu (shishutsu jungaku)* (Ministry of Foreign Affairs, Economic Cooperation Bureau, *The geographical distribution of Japan's flow of resources (net disbursements)*), June 1968.
1969-73: Ministry of Foreign Affairs, Economic Cooperation Bureau, Economic Cooperation Division, *Japan's Economic Cooperation in 1973*, June 1974.
1975:   Gaimushō keizai kyōryokukyoku, *Keizai kyōryoku kankei shiryō*, (Ministry of Foreign Affairs, Economic Cooperation Bureau, *Materials on economic cooperation*), October 1976.
1977:   Ministry of International Trade and Industry, *Flow of Resources from Japan to Developing Countries during 1978*, 31 July 1979.

present an analysis of Japan's position as an aid donor and growing world power, which was intended to foster stable political relations between Japan and the developing countries, while the EPA focused on the implications for Japan's economic prospects. MITI concentrated exclusively on the trade impact of economic cooperation and the MOF paid closest attention to the balance of payments and to the financial burdens imposed on the national budget by foreign aid and trade.

Just as the turn of the decade witnessed notable developments in Japan's aid and more frequent government expressions of aid thinking, so the aid administration expanded. The first yen loans made by Japan called forth new procedures for assessing economic cooperation, including the establishment of an Economic Cooperation Department within the MFA's Economic Affairs Bureau, the first time the Japanese Government had explicitly linked economic cooperation with its diplomatic and political objectives — a marked contrast to the export orientation of the counterpart division in MITI. These changes formalised, but in no way coordinated, Japan's incipient aid effort. They helped to decentralise, not consolidate, the aid administration, a process which continued into the 1960s. These disparities arose notably in 1961 concerning the question of control of the Overseas Economic Cooperation Fund (OECF), a newly established government soft-loan agency. As a result of arguments between MITI, the MFA and the MOF, a compromise solution gave legal responsibility to the EPA, with which it still remained in 1978. The immediate differences then were about bureaucratic influence, but there were underlying arguments over the purpose of the OECF and soft loans in economic cooperation policy, and associated jurisdiction over policy. While they may have presumed to do so, none of the ministries fully represented 'the national interest', which is, in any case, an elusive amalgam of perceptions of what is good for a country. They put forward instead contrasting interpretations of that interest, sometimes in conflict with those ideas held by private enterprise and political groups. The rapid entry of Japan into economic assistance programmes meant that at no stage did there appear a 'government' view of Japan's objectives in undertaking economic cooperation. Any articulated objectives were the product of separate ministry assessments of economic cooperation. Interministerial conflict meant that interests lay more in defending a particular interpretation at home than in formulating government guidelines for a comprehensive aid policy. This, as other writers on donor policy have pointed out, is a well-known phenomenon in donor countries.[23]

1962 saw the end of the substantive development of Japan's aid administration. In that year the MFA, the MOF and MITI all modified their economic cooperation bureaucracies (including a full Economic Cooperation Bureau in the MFA and a Department in MITI), and the second executive agency, the Overseas Technical Cooperation Agency (OTCA), was established under the control of the MFA. These were to be the last significant reforms of the aid bureaucracy until JICA was formed in 1974. By 1963 the lines which were to determine the overall patterns of policy-making from then on had been drawn, the ministries having 'staked their claims' in the different areas of economic cooperation. The four-ministry committee for approving government loans was operating, the MFA had secured majority control in technical cooperation, and responsibility for cooperation with international aid organisations was divided between the MFA and the MOF: Japan's aid administration had fully developed, and bureaucratic interests in economic cooperation had become institutionalised in four ministries. The 1960s were to see the Japanese aid effort become one of the largest among DAC member nations, but the basic structure of her aid administration did not change.

The MFA maintained throughout that economic cooperation was a vital component of foreign policy. It staunchly defended the concentration of aid in Asia and stressed the importance of the region and its economic development to Japan. Nevertheless, it recognised the need for projects in which Japan was involved, both in Asia and elsewhere, to be visible. In its view, Japan's economic and political interests were to go together.[24] As the 1960s passed, Asian recipients received greater priority and references to the natural role for Japan in Asia became commonplace. The MFA was also, via its diplomatic missions, the main channel of communication between the developing countries, international aid organisations and the Japanese Government. A stronger note of respect for the Third World emerged towards the end of the decade, with recognition of its collective influence in the United Nations. The MFA paid more attention to the responsibility of Japan, as an advanced nation and a senior member of the DAC, to give improved aid in order to contribute to the solution of the North-South problem. This was especially so in its public pronouncements, such as the policy speech of the Foreign Minister, Miki Takeo, to the 58th Diet session in 1968.

The two UNCTAD meetings in 1964 and 1968 had a visible impact on MITI thinking. Its single-minded pursuit of the trade effects of aid was arrested in UNCTAD I. A negative Japanese Government speech to the conference brought swift critical reaction from the developing countries and equally rapid moves by the Prime Minister, Ikeda Hayato, to override the Ministers of Foreign Affairs, Finance, and International Trade and Industry and impose a more acceptable policy for representation at Geneva.[25] The 1966 MITI report reflected changed attitudes on the part of the ministry. It revealed a greater awareness of the problems, and the influence, of the LDCs and gave more space to discussion of development *per se*. MITI continued to put its own case, however, and drew attention to Japanese domestic economic problems (such as lower *per capita* income and inadequate social capital) as a reason for a low level of aid, and to the needs of the international economic order. Later MITI statements showed a dual development of ministry ideas. First, they demonstrated wider interest in how Japan fitted into the international, and not only the regional, economy. Renewed suggestions that economic assistance be used for security of resources were one result of this shift in thinking. Secondly, MITI appreciated that the quality of Japan's aid should be upgraded, and a cautious, and rather hackneyed, standard for her capacity to improve performance was proposed. It would be 'to the extent of national capacity'.

The MOF supported similar arguments, but its control of much of aid policy through budgeting made it naturally conservative on the aid question. The ministry gave evidence to the Fiscal Rigidity Study Committee of the Advisory Council on Fiscal Affairs on 25 October 1967 and pointed out that while she recognised the demand for economic cooperation, Japan could not increase the amounts given. It considered that merely maintaining the present ratio of aid to national income would require constant effort. The UNCTAD recommendation for 1 per cent of national income to be given as aid did not have to apply to Japan, since *per capita* income was less than that of Europe and reserves of social capital were poor. It was necessary, when examining new commitments, according to the MOF, to calculate the future fiscal burden of foreign aid already committed. The ministry stressed that any improvement in terms should only be agreed to after properly judging the prospective recipient's needs, and care should be taken to ensure that other recipients did not demand the same terms.[26]

Similar opinions were regularly put forward by other MOF officials, echoing the 'poor fellow my country' arguments about low *per capita* income and social capital accumulation, insufficient reserves of foreign currency and limited experience in relations with developing countries.[27] The MOF's main criterion in assessing aid requests was whether or not the aid recipient really wanted to, and could, develop. Its guiding principle, as expressed by these same officials, was that 'if there are requests for cooperation for promoting self-help, or requests which arise out of efforts at self-improvement, then these should be given precedence'.

The chief concerns of the MOF at the time were, in order of priority, the fiscal situation and the budget impact of aid, the effect of aid on the balance of payments, the prospects for recipient development and the related security of committed funds, and the capacity of the country for self-help. The result was an exceedingly negative approach to the whole foreign aid question by the ministry, one which has largely typified its stance in interministerial discussions on foreign aid to this day.

The Turn of the Decade

The period from 1969 to the end of 1973 marked the height of the Japanese aid effort. While Japan rose to become second-largest donor in the DAC, aid was now indispensable to foreign and domestic economic policy. It was better understood and better integrated into other policy. The concentration in Asia, however, of Japan's aid and the extent of private business representation there created tensions which brought strong reactions against the Japanese presence and methods.

Total Japanese flows increased over fourfold from $1,263.1 million in 1969 to $5,844.2 million in 1973, while ODA rose less quickly from $435.6 million to $1,011 million in the same period. As a percentage of GNP, ODA in fact dropped from 0.26 per cent in 1969 to 0.25 per cent in 1973, although DAC averages fell also (see Table 1.1). Technical cooperation increased from $19 million to $57.2 million but increased only slightly as a percentage of ODA. Contributions to multilateral organisations increased as a percentage of ODA from 22 per cent to 24.3 per cent and remained at slightly above the 23.9 per cent average for major DAC nations in 1973. Increases in OOF and private investment were the most remarkable aspects of Japanese economic cooperation flows in the period, rising from $375.8 million to $1,178.9 million and from $541.7 million to $3,647.5 million respectively. DAC comparisons of aid-giving performance, however, put Japan's grant element of total ODA at 68 per cent in 1969 and 67.9 per cent in 1973, a ranking among DAC members of fifteenth in both years. Geographical distribution (Table 1.2) showed a swing in ODA away from Asia (100 per cent in 1969 to 88.1 per cent in 1973), although the proportion going to Southeast Asia rose (48.5 to 53.8 per cent), as did flows to Africa and Central and South America. Private flows and OOF revealed a remarkable drop in the proportion of resources going to Asia and a large boost in those directed to Central and South America. The DAC continued to pressure Japan to improve her performance. Japan accepted the target of 1 per cent of GNP as development assistance at UNCTAD II in 1968 and agreed to the 0.7 per cent ODA target in 1970, but without a target date. She was at the forefront of international coordination on the untying issue and at UNCTAD III in 1972 announced her intention to untie multilateral contributions in principle. LDC untying was promised from December 1972, and OECF and Eximbank legislation was amended in November 1972 to allow those organisations to make untied loans. Following DAC agreement in June 1974, all loan agreements concluded after 1 January 1975 were in principle LDC-untied.

Ministry attitudes did not really converge greatly. The MFA continued to put the most comprehensive case for economic cooperation and argued that aid should not be given only when foreign exchange was plentiful. The emphasis fell on Japan's standing as an advanced industrial nation and on her international and regional responsibilities. Aid should be considered as one of Japan's basic policy priorities, the ministry maintained, a means to prosperity and welfare both for the developing countries and for Japan. In the opinion of one senior official of the MFA, Japan had to avoid becoming isolated in the interdependent economic order of the 1970s.[28]

An internal MFA document of July 1972 was more frank. It criticised other ministries for their narrow views of foreign aid and tried to explain the myths held about Japanese aid for so long: the practice of tying aid to exports, the negative attitude towards commodity aid and technical assistance and the uninformed ideas about social development in the Third World. The paper stated that although Japan was, in one way, forced to give aid, she should really be giving it to assist LDC development, not to further the policy goals of Japan.[29] In another paper, Councillor Kikuchi of the Economic Cooperation Bureau attacked the use of the words 'economic cooperation', suggesting 'development assistance' (*kaihatsu enjo*) as the more precise and appropriate subject for debate.[30]

MITI still regarded economic cooperation from the viewpoint of international economic policy. Koyama Minoru, Director of MITI's Economic Cooperation Department, wrote that in the 1970s Japan had to change her industrial structure and promote the long-term international division of labour and international economic harmony. MITI's 'aid philosophy', as he called it, was to promote aid which was for the recipient's real benefit, but which more importantly could be turned to Japan's eventual benefit also.[31] The 1971 report put it more abstrusely:

> Japan's relations with the less-developed economies have an importance not seen in relations with other advanced nations. Whether or not the LDC economies can show healthy growth has a serious bearing on our own economy . . . We cannot afford to neglect friendly economic relations with the LDCs. Our position is that Japan's economic cooperation is not simply an international responsibility but an unavoidable requirement for the smooth management of our own economy.[32]

The MOF continued to monitor the effectiveness of the resources Japan was giving and changes in the requirements of the particular developing country. In conjunction with this, it was considered important that LDCs made further efforts to help themselves, although the MOF also stressed that before aid could be sanctioned by the Japanese Government, there must be a 'national consensus' on the need for that aid. Aid budgets could therefore be increased only when this consensus was reached. The MOF used the tactic most effectively in aid budget negotiations when the 'low *per capita* income' and 'weak balance of payments position' arguments of the 1960s became less plausible.

Crisis, Recession and Recovery

The years following the oil crisis brought recession to the Japanese economy, stagnation to the aid programme and, after the middle of the decade, recovery and a new optimism in foreign economic policy. Japan remained one of the two or three largest contributors in the DAC, both in terms of total flows and ODA, although as a percentage of GNP Japan's ODA remained low in comparison with all other donors (except a few) and with the DAC average. The remarkable fall in total flows between 1973 and 1975 and its subsequent recovery to twice the 1973 figure by 1978 was due to the response by Japanese investors to the general economic climate and prospects for overseas investment. Japan's slower performance in ODA up to 1976 was a result of low disbursement of loan funds and a policy of restrained commitment, a fall in grants due partly to reparations being wound down and only small increases in technical cooperation, but ODA doubled between 1976 and 1978. Geographical distribution widened over the period, especially because of large payments to the Middle East following the Yom Kippur War and the Arab oil embargo, and a conscious attempt to improve Japan's relations with countries, such as the least developed nations in Africa, which had played little part in the aid programme until then. Of ODA in 1973, 88.1 per cent went to Asia, 2.6 per cent to Africa and 0.1 per cent to the Middle East, but by 1977 this pattern had altered, so that 60 per cent was directed to Asia, 17.1 per cent to Africa and 12.4 per cent to the Middle East.

Japanese aid in 1978 faced a testing period, after the post-oil-crisis slump in outlays and the gradual return to pre-oil-shock levels of disbursement. The increase in aid budgets fluctuated over the period, with some inconsistency in the weight accorded individual items, although by the fiscal 1979 budget the trend was definitely upwards. Only strong political initiative, however, brought improvements in aid policy. The important reforms to the system were due to firm political pressure on diverse bureaucratic interests: for example, the Japan International Cooperation Agency (JICA) was established in 1974, and the clearer division of OECF and Eximbank financing in June 1975 — giving the OECF control of all ODA loans — was brought about by a well-timed suggestion of the late Minato Tetsurō, a Liberal Democratic Party (LDP) member of the House of Representatives involved in aid affairs. Japan's strong aid relations with ASEAN were reaffirmed during a visit to several Southeast Asian nations in August 1977 by Mr Fukuda, his espousal of the 'Fukuda Doctrine', and his apparently sincere efforts to increase aid disbursements were indicative of resolve at the highest level to improve Japan's image as an aid donor. Japan's promises seemed to be linked to more complex responses to inter-national criticism of her foreign economic policy and this, coupled with more lively domestic discussion of aid issues, assured positive official responses.

Nevertheless, the ministries persisted in their independent views on aid. While all recognised the implications of the oil crisis for the resources rationale of aid policy, there was no move towards a basic government policy on development assistance, but the complexity of the development issue and of the need to realign aid policy to provide aid which suited the economic conditions of the country in question was increasingly appreciated.[33] The MFA developed a more diverse attitude to aid and economic cooperation, on the one hand acting as the most outspoken supporter of the aid programme among the ministries and, on the other, developing an articulate case for Japanese aid policies in the light of the oil crisis and the attendant shift to lower economic growth in Japan. The MFA's first lengthy review of the aid issue and Japanese policies was published in 1978 and closely resembled MITI's annual report in its format. It accepted that there were many factors influencing a donor's economic cooperation policy, but singled out five main considerations for Japan: international economic security, Japan's duty as an economic power, economic self-interest, humanitarian concern and what could be called normal diplomatic necessity. It was, above all else, a very realistic appraisal of Japan's position as a donor, far franker than public statements on aid by other ministries.[34]

MITI continued to regard aid and economic cooperation primarily from a trade perspective. One senior International Trade Policy Bureau official privately admitted that while MITI's attitude might appear 'poor' from the outside, MITI and Japan had to disregard criticism and accept that Japan was a resource-poor trading nation whose domestic and foreign policies could be well served by foreign aid policy. He believed that Japan should be positive and energetic towards aid policy but that the relevant issues were resources, industrial relocation (which applied in the case of Japanese aid to the Asahan hydroelectric power and aluminium refinery project in Indonesia) and exports. It was his opinion that only a ministry, such as MITI, with the power of administrative guidance could effectively manage increasingly complex policy questions such as aid. His views were echoed by MITI's own 1976 review, which pointed out in some detail that the expansion of economic cooperation was first and foremost related to the economic interdependence between Japan and the developing countries and the need to preserve harmonious bilateral trade and investment relations.[35]

Nevertheless, growing trade surpluses and international pressure on the Japanese to curb exports and expand imports made the non-resource trade-generating effects of aid less important to some officials. The other rationales for aid put forward by the MFA in its 1978 review seemed more relevant. The country's diplomatic goals had been well supported by aid in the past in Asia and the Middle East, and the events of 1977-8 strengthened the view that Japan's desire to promote stability in Southeast Asia was expressed in her aid policy towards ASEAN and Indochina — assuming, of course, that those commitments were carried out.

The MOF's emphasis remained different and still conservative. While officials admitted that the MOF had been too harsh in its attitude in the late 1960s and early 1970s, they argued that in 1976 the fiscal situation was still severe and only a foreign aid policy 'appropriate to a low growth economy' was possible. It was important 'to proceed from the standpoint of economic cooperation rather than aid' and to ensure that aid was given most efficiently and in a way which also assisted the Japanese economy. Officials reiterated the consensus argument, claiming that approval of foreign aid by the Japanese people was essential. Only aid which brought benefits to Japan could, in their view, win that sanction.[36]

This unyielding stance was given full play in the domestic argument over implementation of Mr Fukuda's pledge to double ODA within three years. Doubling 1977 ODA disbursement would involve a total of $2,848 million in 1980, but the MOF and MFA disagreed seriously over both the base year and whether the target should be expressed in dollars or yen. While the MFA and MITI preferred yen-denominated ODA based on calendar 1977, the MOF and the EPA wanted dollar-denomination based on calendar 1976, arguing that to double the yen value of Japan's ODA in three years 'would be fiscally impossible'.[37] The compromise adopted was for denomination to be in dollars on a 1977 base. On MFA calculations, the gap between the yen value of a doubled dollar effort (assuming $1 to equal 200 yen) and the doubled yen value of 1977 disbursement would be just under 200,000 million yen, a very substantial discrepancy. 1978 aid figures showed, in fact, that Japan would probably have little trouble in reaching her dollar goal, having gone more than halfway in one year.

A survey of Japan's aid programme in the late 1970s revealed a continuing emphasis on bilateral loans as the core of policy. Loans made up 51.8 per cent of ODA in 1978, and improving loan disbursement was an important goal. While both technical assistance and capital grants formed only a relatively small part of ODA in 1978 (10 per cent and 7.3 per cent respectively), both forms of aid were receiving enlarged budget allocations — bilateral grants up 66.7 per cent in the 1979 budget — and could be expected to rise in gross terms. It was likely that disbursements would improve in the years ahead, as grants and technical assistance were given prominence in the aid-doubling plan, and as some streamlining of disbursement procedures took place. Japan continued to provide about one-third (30.9 per cent) of her ODA in the form of multilateral assistance, in a proportion well above other main DAC donors, though for this reason there was little likelihood of much more ODA going into multilateral aid. Apart from ODA, Japan still provided large amounts of Other Official Flows, mainly in the form of export credits.

In qualitative terms, Japan still lagged behind other main donors in the terms, conditions and grant element of her loans, although the government claimed to be improving these and, in fact, the overall grant element of Japan's ODA increased to 75 per cent in 1978 from 70.2 per cent in 1977. General untying was carried through on about half of loans disbursed, and the government accepted the need to distinguish levels of income between the developing countries and adjust the flow of funds accordingly. One important policy was the attempt to speed up the disbursement of OECF loans, although later chapters will explain some of the difficulties in doing this.

Movements in Japanese policy were influenced by the international aid debate. Even though a sectoral breakdown of her aid showed the projects, in particular those in the industrial or public work and social infrastructure sectors, were preferred, Japan had begun to respond to the needs of agriculture, fisheries, small industry, regional development and the 'basic human needs' programme including education, health, housing and public services. There was still, however, great importance attached to large-scale projects, such as those in Indonesia, Brazil and the Middle East for resources development.

In terms of geographical distribution, Japan's long-standing ties with Asia, especially Southeast Asia and South Korea, meant that aid flows were still directed in the main to that region. This is not to say that other areas were not of relevance — they increasingly stood out in the aid figures — but history, proximity and economic links made Asia the focus. The MFA still regarded stability in Southeast Asia as a necessary condition of Japan's security. Future aid relations with China, first broached in December 1978, and Vietnam, added another highly political dimension to Japan's Asian aid policy which will further complicate the domestic policy debate. In late 1979 Japan was considering a request from China for concessional loans of $5.5 billion, a figure over double Japan's total ODA to all developing countries in 1978. The Japanese appeared likely to accede to considerably less than the Chinese request, but still on a scale rivalling loans to Japan's long-standing leading recipients. Commitments to ASEAN over the period 1977-8 signified that Japan continued strongly to support the development of these vibrant economies, although the most serious question was that of when and how soon Japan could fulfil those pledges, especially those to the incipient ASEAN industrial projects. ASEAN criticism of sluggish Japanese performance and unfulfilled promises indeed typified much developing-country reaction to Japanese policies. The problems of performance demonstrated clearly the patterns that arose where policy responsibility was diffuse and political will inconsistent. Policy initiatives in 1977 and 1978 indicated rethinking of Japan's programme at the highest levels and, in the short term, the achievement of the government's promise to double aid and to improve its terms and conditions will be the real test of the receptivity of the domestic bases of policy to external demand. Performance, however, had its roots not only in ideas and attitudes about aid, but in the structure of the aid system itself, a far more permanent feature of foreign aid than trends in disbursement, shifting percentages or individual bilateral ties. How politics and bureaucracy could become intertwined is the subject of the following chapter.

Notes

1. Five major descriptive studies have been written in English of Japanese foreign aid. The first was a well-balanced critique by John White (*Japanese Aid* (London, Overseas Development Institute, 1964)) and the longest a detailed but incomplete and oversimplified book by Hasegawa Sukehiro (*Japanese Foreign Aid: Policy and Practice* (New York, Praeger, 1975)). An unpublished Ph D dissertation by Edgar C. Harrell focused mainly on budgeting and allocation of aid ('Japan's Postwar Aid Policies', Columbia University, 1973) and a monograph on recent trends in Australian and Japanese aid policy by Nancy Viviani (*Australia and Japan: Approaches to Development Assistance Policy* (Canberra, Australia-Japan Economic Relations Research Project, Australian National University, 1976)) looked mainly at their relations with Indonesia and Papua New Guinea. Finally, there was a perceptive general article by J. Alexander Caldwell ('The Evolution of Japanese Economic Cooperation: 1950-1970' in Harald B. Malmgren, *Pacific Basin Development: The American Interests* (Lexington, Mass., Lexington Books, 1972), pp. 23-60). Martha F. Loutfi also made a cost analysis of Japanese aid in purely economic terms (*The Net Cost of Japanese Foreign Aid* (New York, Praeger, 1973)). There has been no study of the aid bureaucracy in Japan, not even in Japanese, although Japanese works on aid and economic cooperation are numerous. I shall not attempt to summarise that literature here. Hasegawa discusses some of the works in his *Japanese Foreign Aid*, Ch. 1.
2. Reparations to South Vietnam were agreed on in 1960 and quasi-reparations were settled with Thailand in 1962, South Korea and Burma in 1965, Singapore and Malaysia in 1968 and Micronesia in 1972. A small grant said to be in place of reparations was made in 1976 to Mongolia. See Gaimushō keizai kyōryokukyoku, *Keizai kyōryoku kankei shiryō* (Ministry of Foreign Affairs, Economic Cooperation Bureau, *Materials on economic cooperation*) (July 1974), pp. 8-13. Hasegawa briefly discusses the significance of reparations in *Japanese Foreign Aid*, Ch. 4.
3. OECD, *Flow of Financial Resources to less-developed Countries, 1956-1963* (Paris, 1964), p. 142.
4. Tsūshō sangyōshō, *Keizai kyōryoku no genjō to mondaiten* (Ministry of International Trade and Industry, *Economic cooperation: present situation and*

problems) (Tokyo, Tsūshō sangyō chōsakai, 1958). It is hereafter cited as *Economic Cooperation*.

5. See Harrell, 'Japan's Postwar Aid Policies', pp. 35-63, for a discussion of various business initiatives.

6. Gaimushō, *Waga gaikō no kinkyō* (Ministry of Foreign Affairs, *Diplomatic blue book: review of foreign relations*), no. 1, (September 1957), hereafter referred to as *Blue Book*.

7. There was still confusion, however. The 1958 edition, for example, referred to reparations payments to Burma and the Philippines as 'economic aid' *(keizai enjo)* in the table of contents (p. 2) but 'economic cooperation' *(keizai kyōryoku)* in the corresponding text (p. 41).

8. For details of economic diplomacy by Prime Minister Kishi and the Japanese business world, see Harrell, 'Japan's Postwar Aid Policies', Ch. 2; Caldwell, 'The Evolution of Japanese Economic Cooperation', pp. 32-8; and Lawrence Olson, *Japan in Postwar Asia* (London, Pall Mall Press, 1970), Ch. 2. First-hand accounts are also set out in Nihon puranto kyōkai, *Nihon puranto kyōkai jūnenshi* (Japan Consulting Institute, *A ten-year history of the Japan Consulting Institute*) (Tokyo, 1967). This group is an association of machinery exporters. William E. Bryant also discusses private business participation in international economic affairs in more general terms in *Japanese Private Economic Diplomacy: An Analysis of Business-Government Linkages* (New York, Praeger, 1975).

9. See Olson, *Japan in Postwar Asia*, pp. 39 ff. A MITI economic cooperation official at the time, Hayashi Shintarō, noted that there was strong informal pressure from business for the aid to India. See his 'Tai-in enshakkan no haikei to sono igi ('The yen loan to India: its background and significance'), *Ajia mondai*, vol. 7, no. 2 (August 1957), pp. 124-38 at p. 125. Harrell corroborates this in 'Japan's Postwar Aid Policies', p. 41.

10. P.J. Eldridge, *The Politics of Foreign Aid to India* (London, Weidenfeld and Nicolson, 1969), p. 32.

11. OECD, *Flow of Financial Resources to less-developed Countries*, p. 142.

12. Seymour Rubin, *The Conscience of the Rich Nations: The Development Assistance Committee and the Common Aid Effort* (New York, Harper and Row, 1966), p. 70.

13. F.C. Langdon, *Japan's Foreign Policy* (Vancouver, University of British Columbia Press, 1973), p. 87.

14. *Nihon keizai shimbun*, 20 February 1960 (hereafter *Nikkei*).

15. *Nikkei*, 11 July 1961 (evening).

16. Sawaki Masao, in a roundtable discussion 'DAG kaigi no keika to kongo no mondaiten' ('Progress of the DAG meetings and future problems'), *Keizai kyōryoku*, no. 53 (June 1961), pp. 5-13 at pp. 6-7. Similar sentiments were expressed by the Director of the MFA Economic Cooperation Department, Kai Fumihiko, in 'Daigokai kaihatsu enjo gurūpu (DAG) kaigi ni tsuite' ('The Fifth Meeting of the Development Assistance Group (DAG)', *Ajia kyōkaishi* (August 1961), pp. 3-7.

17. Tsūshō sangyōshō keizai kyōryoku seisaku kenkyūkai, *DAC to teikaihatsukoku enjo mondai* (Ministry of International Trade and Industry, Economic Cooperation Policy Research Committee, *DAC and the problem of aid to underdeveloped countries*) (Tokyo, Ajia keizai kenkyūjo, 1966), pp. 1-6.

18. *Nikkei*, 24 October 1961.

19. Leon Hollerman, *Japan's Dependence on the World Economy* (Princeton, Princeton University Press, 1967), especially p. 204, note 26.

20. For example, DAC used the category of 'official assistance' from 1961, separating government and private flows of capital, but MITI was slow to respond,

not presenting its own aid statistics with government and private flows separately listed until the 1968 edition of *Economic Cooperation*. Editions from 1962, however, did carry tables provided by DAC showing government and private flows from all members including Japan.

21. For a discussion of the concept of 'net flow of resources', see OECD, *Development Assistance: Efforts and Policies of the Members of the Development Assistance Committee, 1969 Review* (Paris, 1969), pp. 239-41 (hereafter DAC, *Review*). DAC defines Official Development Assistance as 'all flows to less-developed countries and multilateral institutions provided by official agencies, including state and local governments, or by their executive agencies, which meet the following tests: (a) they are administered with the promotion of the economic development and welfare of the developing countries as their main objective; and (b) their financial terms are intended to be concessional in character'. Other Official Flows include '(a) official bilateral transactions which are not concessional or which, even though they have concessional elements, are primarily export-facilitating in purpose; (b) the net acquisition by governments and central monetary institutions of securities issued by multilateral development banks at market terms. Rediscounting of trade instruments by central monetary authorities is not considered as an official flow.' By this definition, OOF includes Japanese Export-Import Bank suppliers' credits and OECF suppliers' credits, recorded until 1968 as private transactions. See DAC, *Review* (1969), pp. 241-3.

22. The Export-Import Bank of Japan (Economic Research Department), *Japanese Private Investments Abroad (A Summary of Third Questionnaire Survey)* (1972), p. 6.

23. There is some literature describing this phenomenon, of which Judith Tendler's book is the most telling. See her *Inside Foreign Aid* (London, Johns Hopkins University Press, 1975). On Britain, Judith Hart tells a practitioner's tale, in *Aid and Liberation* (London, Gollancz, 1973), especially pp. 179-87, as do Nancy Viviani and Peter Wilenski, *The Australian Development Assistance Agency: a post-mortem report* (Brisbane, RIPA National Monographs 3, 1978).

24. *Blue Book* (1962), p. 226. According to this, projects had to be *senden kōka no takai*, 'of a high public relations value'.

25. Langdon discusses the incident on p. 92 of *Japan's Foreign Policy*. Olson notes how Japan learnt 'a sober lesson' (*Japan in Postwar Asia*, pp. 142-4) and Harrell refers to it as 'a turning point in Japan's economic diplomacy toward Asia' ('Japan's Postwar Aid Policies', p. 107).

26. Kawashima, 'Enjo wa naze suru ka: zaisei kōchokuka to keizai kyōryoku' ('Why do we give aid? Fiscal rigidification and economic cooperation'), *Keizai to gaikō*, no. 522 (December 1967), pp. 9-11 at p. 9.

27. For example, Kaya Akira, 'Teikaihatsukoku enjo no shomondai' ('The problems of aid to underdeveloped countries'), *Fainansu*, vol. 4, no. 1 (April 1968), pp. 26-31, and Fujii Yasuhisa, 'Kōshinkoku enjo no genjō to mondaiten' ('The present situation and problems of aid to backward countries'), *Fainansu*, vol. 4, no. 7 (October 1968), pp. 43-9.

28. Motono Moriyuki (Councillor, Economic Cooperation Bureau, MFA) in a roundtable discussion 'Keizai kyōryoku no konnichiteki igi' ('The significance of economic cooperation today'), *Kokusai kaihatsu jānaru*, 5 November 1972, pp. 2-18 at p. 3.

29. Gaimushō keizai kyōryokukyoku, *Kongo no taigai enjo ni kansuru jakkan no teigen* (MFA, Economic Cooperation Bureau, *Several proposals for future overseas aid*), 20 July 1972.

30. Kikuchi Kiyoaki, 'Keizai kyōryoku no dōkō ni tsuite' ('Trends in economic cooperation'), speech to Kaigai denki tsūshin kyōryokukai (Overseas Telecommunications Cooperation Association), 17 May 1973.

31. *Kokusai kaihatsu jānaru*, 5 November 1972, pp. 4-5.

32. *Economic Cooperation* (1971), pp. 116-17.

33. See Taigai keizai kyōryoku shingikai, *Kongo no kaihatsu kyōryoku no suishin ni tsuite* (Advisory Council on Overseas Economic Cooperation, *On the promotion of future development cooperation*), 18 August 1975.

34. Gaimushō keizai kyōryokukyokuchō-hen, *Keizai kyōryoku no genkyō to tembō: namboku mondai to kaihatsu enjo* (Ministry of Foreign Affairs Economic Cooperation Bureau Director General (ed.), *Economic cooperation: present situation and prospects: the North-South problem and development assistance*) (Tokyo, Kokusai kyōryoku suishin kyōkai, 1978), pp. 355-60.

35. *Economic Cooperation* (1976), pp. 182-6.

36. Interviews with MOF officials, June-July 1976.

37. See *Asian Wall Street Journal*, 4 July 1978, and *Yomiuri shimbun*, 10 June 1978.

2 'SCRAP AND BUILD': THE ORIGINS OF JICA

As the Japanese aid programme expanded rapidly after 1955, most administrative growth in the aid policy area was determined by ministries pursuing their broader policy interests. Because of a rigid Japanese administrative tradition and persisting conflicts about the purposes of aid, the aid bureaucracy was fragmented and administrative change failed to keep pace with aid flows. Power over policy was divided and no natural policy leader emerged from among the ministries.

Why this was so reveals much about aid policy-making. John White maintains that the choice of one type of aid administrative structure rather than another 'is usually the outcome of historical accident combined with the administrative conventions of the country concerned'.[1] In Japan's case, however, accident and convention do not fully explain administrative change over twenty years. The unsteady relationship between aid and politics was decisive in linking motivations and mechanisms; political pressures were a significant, if only negative, influence. Arguments over the purposes of aid led to competition between ministries for control of policy.

This tussle is illustrated well by the way JICA came about. The agency was formally established on 1 August 1974 to manage Japan's technical aid programme, some development funding and to assist emigration. Yet JICA's final shape was rather unexpected, the result of a hurried ministerial attempt to resolve an impasse over rival budget proposals. It is a story which epitomises the relationship between aid and diffuse policy currents within the Japanese Government. It illustrates the shifting balance of interests in aid policy, the ordering imposed by budget procedures and the legislative process, and the separation of the domestic politics of aid from ideas of economic development. From several conflicting perceptions of the purpose of the agency there emerged a body which satisfied few, if any, of those original goals. Haste typified the behaviour of politicians and officials alike. Bureaucratic politics directed the debate and narrowed the final options, but proved unable ultimately to resolve an issue which impinged on the goals of aid sections in different ministries. In the final outcome, discussion centred on the merits of combining existing agencies rather than on the benefits of JICA as such.

Bureaucratic Change and a Central Aid Agency

How bureaucracies grow and change depends a great deal on the conflict of ideas. The development of new bureaucracies resulting from
the 'constant flux in the nature of policy space'[2] is a familiar theme:
existing bureaus interact with the changing environment and strive to
preserve independence in pluralist systems responding to shifting social
demands.[3] Herbert Simon showed in 1953 how environmental forces
interacted with conflicting conceptions of the tasks of the United
States Economic Cooperation Administration to affect that agency's
development. Processes of organisational change, he argued, were learn-
ing processes where 'environmental forces mold organizations through
the mediation of human minds'. Changing the organisation meant
changing programme goals.[4] Throughout the history of the Japanese
aid programme — as Chapter 1 revealed — contrasting perceptions of aid
were the stimulus for change in aid management structures. These were
domestic attitudes, mainly a product of ministry approaches to problems, although moulded by politics also.

Despite the varied forms of aid administration in donor countries,
there was a natural tendency for dispersal of control over aid programmes, even in centralised systems. This was because 'control' related
not simply to political responsibility and corresponding administrative
structures, but to bureaucratic processes and their political effects.
During the growth of aid bureaucracies, strong political direction was
an essential, but not always successful, counterbalance to the disaggre-
gating tendencies of accident and convention, but such forces never
combined to produce a central aid agency in Japan. JICA is perhaps the
closest in form, although it was at no stage intended to control the
whole aid programme. There was never a well-developed movement or
political impetus for reform able to overcome the weight of the established bureaucracy. Ideas for change appealed always to notions of
'administrative unity' which ministries influential in aid policy preferred to avoid.

Calls for a central aid agency usually followed criticism of the
administration and its interminable internal disputes. The first came in
August 1957 from a group of Liberal Democratic Party (LDP) members, then in 1958 from the Federation of Economic Organisations and
the LDP's special aid committee. In 1964 the government's Commission
on Administrative Reform suggested changes and in April 1967 the
Foreign Minister, Miki Takeo, announced that an agency was being considered. He repeated this call for an agency in July 1970 but ministerial

speeches were not enough. It was, in fact, in the interest of the MFA to opt for the *status quo* and retain the control it already had over aid.

Nevertheless, the idea of an agency retained its glamour. In February 1968 the Japan Committee for Economic Development, one of the three largest economic organisations in Japan, argued that an agency was essential for consistency and flexibility in policy. This was echoed three years later in an interim (but *not* in the final) report of the government's Aid Advisory Council, while the LDP's special committee on aid sought the merger of the OECF and the Overseas Technical Cooperation Agency (OTCA). Pressure for change came not only from within Japan: the DAC, in its annual examination of Japan's effort, often referred to the need for administrative reforms.

By 1978 the momentum was weaker. Officials were unclear about how to resolve the clash of ministry interests and few were able to see how a start could be made. Many accepted the need to centralise, but could not say from where the political impetus would come, if indeed there were any. Official publications shied away from the subject of a central aid agency. The MFA's 1978 report on aid made passing reference to the problems of a dispersed aid administration, merely cautioning the need for 'maintenance of a unified foreign policy'. At the highest political levels there were continuing doubts about how independent any aid ministry might be, because of the entrenched administration. The then Prime Minister Fukuda put this view during a meeting of the Advisory Council on Overseas Economic Cooperation on 24 November 1978.

In the past, political support for aid policy was sporadic and given only for specific ministry interests. It was therefore crucial in *controlling* administrative change and *preventing* comprehensive reform. This was not difficult to achieve, since proponents of change did not stress the beneficial effects of a restructured system or the developmental impact of donor administration; debate centred instead on the vague notion of 'unifying' policy, so that bureaucratic inertia easily smothered piecemeal reform proposals. There was no sense of immediacy about reform until the early 1970s, when international conditions began to favour a more coordinated approach to aid management, and at least one ministry (the Ministry of Agriculture and Forestry) linked its own policy strategies with administrative solutions to Japan's aid performance. We turn now to see the effects of this initiative.

Bureaucratic Conflict and Budgeting

In August 1972 the Ministry of Agriculture and Forestry (MAF) included in its request to the Ministry of Finance for the 1973 budget a proposal to establish a new agency for the development of agriculture in the Third World.[5] The request was fairly substantial (12 billion yen) and was prompted by concern within the ministry on two issues: agricultural import policy and MAF influence in the domestic aid administration, which was restricted by its specialised, largely domestic orientation, its commitment to a viable Japanese agricultural sector and a food policy consistent with that goal.

Those sections of the ministry responsible for agricultural trade were, by the nature of their work, less firmly wedded to these premisses. They clearly understood the effect on Japan's food supplies of trends in international agriculture, and recognised the need to increase assistance to the primary sector of developing countries. Japan depended heavily on imports for many of its staple foods – cereals (except rice), soya beans, sugar and so on – and in 1972 the developing countries provided 52.2 per cent of her primary produce imports. These MAF officials hoped that policies of overseas agricultural development might help diversify sources of agricultural commodities, for a few developed countries, especially the United States, Canada and Australia, provided the bulk of staple food imports. Wheat was imported from the US, Canada and Australia, maize from the US, Thailand and South Africa, soya beans from the US, sugar from Australia, Cuba and South Africa. Irrespective of this dependence, Japan's performance in agricultural development cooperation at the time was poor when compared with assistance to the industrial sector in the LDCs, and until the late 1960s agricultural aid was important neither in aid policy nor food policy. In 1973 only 4.1 per cent of bilateral project aid went to the agricultural sector compared with 69 per cent to industrial, energy, transport and communications projects.[6] The existing Overseas Technical Cooperation Agency was ill equipped to provide agricultural assistance services and passed on much of this work to the MAF to implement, which proved a complex and unsatisfactory procedure.

Internal administrative questions were also prominent. Within the MAF the International Cooperation Division of the Economic Affairs Bureau was, as one commentator put it, a 'special section', out of place within a predominantly inward-looking ministry.[7] If policies were to be pushed forward, it was realised, aid divisions of the MAF needed more power in the bureaucracy than their status within the ministry afforded them. Officials of the division certainly aspired to this, but the MAF had no power base in the national aid administration. The MAF was not included in the four-ministry group which made decisions on loans policy and its task in technical cooperation was only secondary, mainly to provide specialist staff for agricultural technical aid projects, although it also ran its own training programmes.

The MAF, however, had a rival in its attempt to create a new agency, for MITI had requested funds from the 1973 budget for a Small and Medium Industry Overseas Investment Agency. Because the two proposals were rather similar, considerable difficulties were expected in securing MOF approval. The MAF plan presented problems in the balance between agricultural and forestry projects, for emphasis on the latter was likely to involve the government in schemes too obviously designed out of domestic self-interest. The MITI plan was not envisaged at any stage to be wholly government-operated but was to rely on cooperation from private enterprise.

Initially, the MOF adopted a negative approach. Aid administration was, it asserted, a problem beyond the scope of the MAF's responsibility which demanded cross-ministry solutions. It suggested rather that the MAF examine its proposal more carefully. Thirty million yen was allocated for further research into the scheme,[8] and the MAF foreshadowed a request for the following year's budget. It set up a coordinating study group within its Economic Affairs Bureau in June 1973, and seven missions were sent abroad to research trends in agricultural production and food demand.

The MAF regarded movements in world food production and demand as being of importance for national security because of their effect on Japan's long-term food imports and, as Fred Sanderson quite rightly pointed out, 'Japanese concern over the country's growing dependence on food imports reached new heights under the impact of events commonly described as the "world food crisis" beginning in 1972'.[9] The Planning Division of the Minister's Secretariat conducted in 1972-3 its own study of the food problem, which had been highlighted in the summer of 1972 when the Soviet Union embarked on a programme of buying up grain stocks. Until the 'food crisis' of 1972-4, when stocks fell, demand suddenly rose and export prices of cereals soared, Japanese agricultural cooperation consisted mainly of assistance with rice cultivation and with projects which resembled rather closely the 'development import' ideas current ten years earlier in MITI circles and revived in the late 1960s.[10] Opposition parties had attacked the policy as having only a minor effect on the stability of Japan's food imports, and agricultural pressure groups predictably opposed it as a threat to the incomes of Japan's rice farmers. The international situation which emerged in 1972 gave renewed force to the argument put forward by the MAF's International Cooperation Division that it was important to develop overseas agricultural projects. This was not inconsistent with another strand of MAF thinking that Japan should be aiming for greater self-sufficiency in foodstuffs, expressed in MAF projections of food supply and demand over the 1960s and 1970s. Self-sufficiency by no means implied eliminating imports of foods which were already mainly imported, but rather improving domestic production.[11] In particular, imported feedstuffs would be the basis for the projected shift to livestock production, for example. Overseas agricultural projects were seen, in such times of crisis, as an attractive policy option. Thinking in the ministry was supported, ironically enough, by the US Government's surprise announcement on 14 June 1973 that exports of cereal products would be temporarily restricted. This involved a 50 per cent cut in contracted export quantities of soya beans (an important source of protein for the Japanese) and the vulnerability of Japan's essential food supplies suddenly became apparent to all.

The soya bean 'shock' was perfectly timed to the MAF's advantage, coming only two months before ministry requests for the 1974 budget were due to be made. Greater public awareness of the food issue gave legitimacy to agricultural development proposals, and serious lobbying by 'sponsors' of MAF ideas and intense political involvement replaced the apathy of 1972. Politicians weighed in to support the MAF. Kuraishi Tadao, a former Minister of Agriculture and Chairman of the LDP's main policy development organ, the Policy Affairs Research Council (PARC), called for an agricultural cooperation agency in a paper presented to the party's Special Committee on Overseas Economic Cooperation, and LDP Secretary-General, Hashimoto Tomisaburō, was also quoted as suggesting that the MAF should consolidate its administrative position in agricultural cooperation.[12] Such support to some extent countered expected opposition from the MFA and the Administrative Management Agency (AMA) to the very idea of agricultural cooperation. Both had strongly argued against the concept a year earlier, but their objections in 1973 were to be instead over the emphasis and organisational implications of the policy.

A Committee on Overseas Agricultural Cooperation was set up in June 1973 by Tokonami Tokuji, an LDP member of the House of Representatives (representing a rural Kagoshima constituency) and

Chairman of the Agricultural Sub-committee of the LDP Special
Committee on Overseas Economic Cooperation, paralleling the study
group in the Economic Affairs Bureau of the MAF. At the end of
August the Tokonami Committee produced a proposal for the interministry committee on agricultural cooperation and an Overseas
Agricultural Development Cooperation Agency, which was to include
the Japan Emigration Service (JEMIS). It would concentrate on agricultural cooperation with the Latin American countries and be
supervised jointly by the MFA and the MAF. Minato Tetsurō, another
LDP member actively interested in aid, believed that this proposal
impressed upon many within the LDP that agricultural cooperation
should be taken more seriously.[13]

The report from the study group within the MAF's Economic
Affairs Bureau led to a 40 billion yen request in the 1974 budget (over
three times the size of the previous year's request) for a combined
government-private agency to 'stabilise' Japan's food imports. Unfortunately, the emphasis was too obviously on the benefits to Japan
rather than to the developing country. This gave grounds for the MFA
to oppose the proposal and led eventually to the displacement of
agricultural development as the agency's primary function.

Partly out of pique, MITI responded quickly to the MAF request by
seeking funds for an Overseas Trade Development Cooperation
Corporation, to finance industrial and resources projects in developing
countries. It was to absorb the Japan Overseas Development Corporation (JODC), a body set up in 1970 to assist private firms in developing
Asian resources. While MITI's reaction was an attempt to expand
alongside the MAF, to keep up with its bureaucratic neighbour, the
plan contained elements of that rejected by the MOF the year before.
Acording to Moriyama Shingo, then Director of MITI's Economic
Cooperation Department, writing in *Kokusai kaihatsu jānaru* of 5
October 1973, it was fully consistent with MITI policies of industrial
relocation, international division of labour and support for private
initiative in resource development.

The proposals brought immediate responses from other ministries.
The EPA was against both, for as administrator of the Overseas Economic Cooperation Fund (OECF) it was concerned about the impact of
the proposed agencies on the scope of OECF lending. The MFA also
opposed the two plans but on wider grounds than in 1972. It objected
on the basis of administrative feasibility, policy coordination and
principles of development policy. As the ministry in charge of the
Overseas Technical Cooperation Agency (OTCA), it saw no need for a

new agency, since it believed that technical cooperation was already well managed, and that neither plan was capable of properly coordinating technical and financial aid. It likened the development import content of both to 'resources plunder' which it claimed was against Japan's long-term interests.[14] The MFA's attention was not entirely upon development issues, however, and as in 1972 it obviously desired to preserve its own authority within the aid administration. To that end, official MFA policy regarded the new agency as unnecessary, but its fallback position was that, if inevitable, the agency should be as comprehensive as possible and not restricted to one sector of aid policy such as agriculture. Only this attitude could ensure that the MFA's organisational interests were protected.

Domestic agricultural groups were not all convinced of the worth of the MAF request. Within the ministry itself debate was active over whether the emphasis would be placed on development import or on development cooperation, for recipient benefits featured more prominently in the latter. To many, such as those in the Agricultural Structure Improvement Bureau, development import was a positive course to follow (given their concern for the future of Japan's primary industry structure) whereas it was only in mid-1973 that other sections, such as the Animal Industry Bureau and the Forestry Agency, came to consider either approach as useful to Japan's agricultural policy. Events made it clear to them that diversifying the sources of stock feed and timber imports could assist Japan.[15] An Overseas Agricultural and Forestry Development Policy Group (brought together in the Minister's Secretariat with officials from the secretariat and from the Economic Affairs Bureau), especially its Director, Kawamura Kōichi, with support from the International Cooperation Division of the Economic Affairs Bureau and from its Director, Ashikaga Tomomi, succeeded in shifting the emphasis of the proposal from 'import' to 'cooperation'. It did this to show that, in contrast to MITI, its ideas approached the development solutions sought by the LDCs themselves, without jeopardising Japan's long-term food import structure or the domestic industry. Making this distinction was vital in terms of the domestic political balance. By moving the main theme of the argument from pursuit of Japan's interests to the promotion of LDC goals the policy group quietened MFA opposition and won over many of its opponents in and around the ministry.

These efforts of MAF officials to turn the debate to their advantage was bolstered by a fortuitous turn in political events. This came as the process of bureaucratic politics reached the difficult budget negotiation stage, in which politicians and parties were to become involved. While this would open up the debate to wider political forces, the strict time-table of the budget process imposed its own organisational order and allowed the MAF's supporters in Cabinet to protect its interests.

As a result of the sudden death of the Finance Minister, Aichi Kiichi, the Prime Minister, Tanaka Kakuei, reshuffled his second Cabinet on 25 November 1973. Three rural Diet members, among the most power-ful men in the LDP and all of the same Fukuda faction, took over portfolios directly associated with the agency question: Fukuda Takeo became the Minister for Finance, Kuraishi Tadao Minister for Agricul-ture and Forestry, and Hori Shigeo the Director for the AMA.[16]

Lobbying continued as part of the normal budget process through November and the first half of December, although it merely hardened the positions of the ministries. The MAF put its case to the LDP Policy Affairs Research Council's Agricultural and Forestry Division and the Special Investigation Committee for Basic Policy on Agriculture, Fores-try and Fisheries, with the former Minister of Agriculture and Forestry, Sakurauchi Tadao, acting as coordinator. MFA and MITI lobbied their counterpart divisions. The backing of ministers, LDP policy committees and business groups was sought, but private enterprise was not obviously united, and many larger firms (especially trading companies) stood to gain from both schemes. At the time, the MAF plan was perhaps more politically acceptable than MITI's, again helped greatly by chance. The Middle East War had begun in October 1973 and once more Japan was made painfully aware of the instability of her resource imports. It became 'fashionable' to promote agricultural development and the political edge enjoyed by the MAF was reflected in the discrepancy between its own ardent attempts to ensure support and a rather tepid approach by MITI.

Because of the controversy, the outcome of the budget requests of the two ministries was not surprising. In the MOF draft budget of 22 December 1973 neither ministry was allocated funds for an agency and a decision was held over until the 'revival negotiations' between the MOF, ministries and ministers had taken place. Deferral was a normal procedure for sensitive political requests. While it was evident that a political decision was required, other non-political factors had influ-enced the MOF. Finance Minister Fukuda's policy for the 1974 budget was, in response to the effects of the oil crisis, highly deflationary and an enlarged administration was not a priority in that fiscal climate. At Fukuda's insistence, the MOF draft recorded a modest growth of 19.7 per cent over fiscal 1973, a reduction, as Campbell explains, 'of 2.7

percentage points of growth from the figure regarded by the specialists in the Budget Bureau as the bare minimum less than a month earlier'.[17] In addition, the AMA's 'scrap and build' policy required the abolition or absorption of existing agencies when new ones were created, to prevent an ever-expanding bureaucracy. Not surprisingly, however, none of the ministries wanted to relinquish any of the agencies in their control to form a new conglomerate aid organisation. The MAF had further upset calculations by requesting a second agency in the 1974 budget, an agricultural land development body which required another existing agency to be 'scrapped'.

Bureaucratic politics, therefore, brought the issue to a political decision, although fate and serious policy motives on the part of some MAF officials were also instrumental. The pursuit of policy questions through the budget process was normal bureaucratic procedure, but one where ministerial rivalries went unchecked and resolution was impossible in policy terms. Organisational process, in the form of budgeting, channelled ministry arguments into a strict timetable. For this reason, debate turned to the immediate task of success in bureaucratic and budgeting terms, and the purposes of the proposed agency became a secondary problem.

A Politicians' Decision

The decision to establish the new Japan International Cooperation Agency was made formally at Budget Cabinet on the morning of Saturday 29 December 1973, after agreement between the Finance Minister and LDP leaders (Chairman of the Executive Council, Party Secretary-General and Chairman of the PARC) in the final budget negotiations on the night of 28 December. In its last week the budget process attracted the close personal attention of the Prime Minister to the aid question and involved heated negotiations between politicians, senior bureaucrats and budgeting officials, in which the claims of MITI, the MAF and the MFA were eventually accommodated in a considerably narrowed set of options.

The first step in breaking the impasse between ministry positions came when Minato Tetsurō, in a proposal released on 21 December (but based on an earlier one prepared in September), suggested the establishment of both a Policy Office in the Prime Minister's Office to coordinate aid policy and an Overseas Economic Development Corporation. The corporation was to be formed from three existing agencies, the emigration assistance body JEMIS (under MFA direction), MITI's JODC and the Overseas Agricultural Development Fund (a small financing organisation under the MAF), with tripartite control. It was a curious document for Minato to produce, for it left several administrative problems (such as the status of the OTCA) unresolved. The proposal had been written in close consultation with the Minister for Agriculture, Kuraishi, and suggested an MAF initiative to force a favourable compromise.[18] Other LDP sources corroborated this by claiming that Minato wrote the paper 'for the Party and the ministries'.[19]

Minato pursued the issue, and paid a private visit to the Prime Minister on Sunday 23 December. These talks were a turning-point in the already drawn-out discussions, for as a result Tanaka issued a directive to government officials on Tuesday 25 December to establish new machinery for economic cooperation.[20] Minato asked the Prime Minister for three things: a 'control tower' for aid in the form of a responsible minister; a stronger aid advisory council; and an agency.

There was some disagreement between the two men over these requests. Minato thought that, because the aid administration was already so complicated, an incumbent minister could be given the added aid responsibility. Tanaka, however, was adamant in preferring a new ministerial post. In regard to the proposed agency, Minato now hoped to build a conglomerate organisation from not only the JEMIS and the JODC, but also the OTCA and even the OECF and the Export-Import Bank (Eximbank). This was a far broader concept than his proposal of 21 December and reflected something of the MFA attitude. Both Tanaka and Minato realised, however, the administrative conflicts which would arise from this, and Minato yielded to a simplified arrangement. The Tanaka memo to officials of the 25th incorporated all three of Minato's suggestions, with Tanaka's amendments.

In the meantime, the budget process continued as normal. 'Revival negotiations' (between ministries and the MOF about initial MOF allocations) were concluded at desk-officer level by Monday 24 December, while the policy committees of the LDP continued to conduct hearings with ministries. On the morning of Tuesday 25 December, the Prime Minister instructed the Councillors' Office of the Cabinet Secretariat (through the Director of the Cabinet Legislation Bureau, Yoshikuni Ichirō) to prepare detailed proposals in accordance with his memo. In the evening, a meeting of the LDP Policy Affairs Research Council (PARC) Deliberation Commission framed party policy for the 1974 budget and at 11.00 a.m. the following day a special meeting of the LDP Executive Council was held to ratify the previous evening's eight-point budget programme. The council decided among other things that the party should recommend the merger of the MITI and MAF agency plans, complete ministerial discussions by the 28th and have the budget approved by Cabinet on the 29th.

The focus was, at this point, on two issues: the form of the new aid agency and the status of the proposed new minister for economic cooperation. The *Yomiuri shimbun* of the evening of Wednesday 26 December reported prematurely (and inaccurately) that the LDP had decided to set up a new ministerial post for economic cooperation and a new agency, after agreement had been reached between Ōhira, Fukuda and Tanaka. Certainly, the Prime Minister met his two top ministers separately on the morning of the Tuesday and no doubt mentioned his plans to them then, but discussion was brief. Ōhira opposed the plan for a minister, warning of the dangers of 'double diplomacy' and interference with his own portfolio. Fukuda's initial attitude appeared in his statement with Hori on the following day that, in view of the difficult budget situation, no new government agencies would be established except the Housing Development Corporation,[21] a priority request first raised in 1972 and pushed by LDP Secretary-General and Tanaka faction leader Hashimoto Tomisaburō.

These were early negotiating stances. The Finance Minister, Fukuda, again met the Prime Minister on the morning of Thursday 27 December but would not accede to the creation of an agency, stating that budget policy demanded a strengthening of aid in qualitative rather than in quantitative terms. The two also differed on details of the new Cabinet position, even though they agreed in principle that it should be created. Tanaka preferred a post with full powers and a complete staff but Fukuda suggested that, for mobility and impact, a 'roving' minister with only a small staff would be better. The minister would be without portfolio, to avoid administrative complications. Fukuda hoped to see the minister 'complement the already over-worked Foreign Minister'.[22]

Tanaka saw the Foreign Minister, Ōhira, on the afternoon of the same day but first demonstrated his own firm resolve to wind up these initiatives in the aid area. He instructed the Director of the Cabinet Legislation Bureau to prepare legislation for a ministerial post, although Yoshikuni reportedly had doubts about both proposals, and then recorded an interview for NHK, the national radio and television network, to be broadcast at 8.00 p.m. that night, in which he announced the government's intention to establish an agency for overseas food and resources development. He did not mention his plan for a minister for economic cooperation. Meanwhile, Suzuki Zenkō, Chairman of the LDP Executive Council and of Ōhira's closest political associates, expressed reservations about the appointment of a minister, and the MFA came out with a fresh scheme to shelve that proposal but to create an agency combining both the MAF and MITI ideas, and incorporating JEMIS. While the *Yomiuri* claimed that this gained concessions for the MFA, it would seem rather to represent a weakening of its stance and an expression of its desire to negotiate on the minister-agency questions, perhaps to avoid the former by agreeing to the latter.

Tanaka met at 1.00 p.m. on Thursday 27 December with Ōhira and several MFA officials, who were to discuss Tanaka's forthcoming trip to Southeast Asia (itself later notable for the riots which accompanied Tanaka's visit, especially in Jakarta). The meeting became difficult when Ōhira demanded retraction of the plan for an economic cooperation minister and even threatened resignation. He warned Tanaka against moving too quickly, and said that if a ministerial position were created, the MFA would provide no staff, 'except a secretary and a girl to make the tea', and that the minister would have to be subordinate to the Foreign Minister. Tanaka apparently reacted strongly to this firm MFA stance, and the meeting proved inconclusive.[23]

Newspapers on the morning of Friday 28 December reported that on the previous day the government had completed arrangements for the minister. This referred to guidelines drawn up by the Cabinet Secretariat, which recommended that the minister should have neither a ministry nor the power to negotiate with foreign governments. He would chair an Economic Cooperation Committee which would be set up in the Prime Minister's Office, absorbing the Advisory Council on Overseas Economic Cooperation. This format clearly separated the question of the minister from the agency proposal and left room for the MFA to accept one without the other.

A meeting of senior ministers after Cabinet that morning made the final decision. The Prime Minister, Fukuda, Ōhira, Hori and Nikaidō Susumu, the Chief Cabinet Secretary, were present, but Kuraishi and Nakasone were notably absent. Despite opposition from Ōhira, it was agreed to establish a ministerial post, but without portfolio and with only a few assistants. The committee idea was rejected, but an economic cooperation agency to be called *Kokusai kyōryoku jigyōdan* was to be created. In a vague five-point directive, later to be nicknamed the *Gokajō no goseimon* after the Charter Oath of Emperor Meiji issued on his assuming power in 1868, it was agreed:

1. that the OTCA and the JEMIS would be the foundation of the agency;
2. that the Foreign Minister would supervise, while the Ministers for Agriculture and Forestry and International Trade and Industry would have joint responsibility with the MFA in relevant areas;
3. that the JODC and sections of the Overseas Agricultural Development Fund would be included;
4. that the present structure of the OTCA would be retained, with new departments for agriculture, mining and industry, and emigration;
5. that a law would be enacted along the lines of the OTCA Law.[24]

This agreement was ratified as a budget item in talks between the Finance Minister and LDP leaders on the evening of Friday 28 December and approved in Budget Cabinet on the morning of Saturday 29 December. The agency was allocated by the MOF a surprisingly large sum of 5 billion yen, probably from reserve funds held for late settlement of sensitive budget items, plus moneys already set aside for the OTCA and the JEMIS.[25]

The Decision: Two Issues

The week of political bargaining which forced a hurried compromise left no room for considering the implications for future policy. Scant regard was paid to solving the policy conflicts between the MAF and MITI, or to the effect of proposals for overseas resources development on technical cooperation policy. The decision to abolish the OTCA, a reasonably successful independent agency, came at a very late stage and was not a serious attempt to improve Japanese technical assistance administration. Even though the AMA was engaged at the time on a report criticising many technical cooperation procedures,[26] the MFA thought that the OTCA was performing its duties quite well.

Including the OTCA in the proposed agency was not raised seriously until Minato's private meeting with Tanaka on 23 December. Until then, the MAF and the MFA had agreed that the JEMIS would be provided as the *zabuton*, or 'cushion' for any 'scrap and build'. There were sound policy reasons for this: it gave scope for extending the useful link established between emigration and agricultural development in, for example, South America, as Okayasu Makoto, Director of the Economic Affairs Bureau of MAF, explained in evidence given on the JICA bill to the Foreign Affairs Committee of the House of

Representatives on 8 April 1974. However, since the MFA had always considered that to be most effective any new organisation should be comprehensive, it recognised that the OTCA would benefit from being incorporated in a larger body, although the possibility was, until late in the process, remote. The Tanaka memo of 25 December (directing the abolition of both the OTCA and the JEMIS) surprised the MFA but, at an emergency meeting of senior officials, a representative of the Economic Cooperation Bureau suggested that it provided a good opportunity to enhance the role of the OTCA by having it 'scrapped' instead of the JEMIS. Appeals were made to Ōhira by MFA emigration and economic cooperation officials, but the minister could not convince Hori, Director of the AMA, that both should not be abolished.

The OTCA was a victim of political circumstance and lack of foresight on the part of MFA officials. They grasped too late the chance to see the OTCA used constructively in the agency, over which the MFA gained majority control only by a late tactical about-face. In this instance politics ran ahead of bureaucratic manoeuvrings. In the impasse between ministerial rivalry and fiscal difficulties, participants were inward-looking and defensive, and the Prime Minister's aggressive use of the scheme for a new minister to soften MFA opposition to a new agency proved the most positive feature of the final negotiations.

This leads us to ask why Tanaka pressed so strongly for a minister for economic cooperation even though other LDP leaders firmly opposed it. His apparent lack of concern for the details was overshadowed by his personal enthusiasm and impatience for a quick decision. One newspaper suggested that it was one of his few personal directives for the 1974 budget, and Hori Shigeo was quoted as saying that 'it is not an administrative problem but one of the Prime Minister's own ideas'.[27] Tanaka's motives were, however, not obvious. He certainly seemed to favour a new coordinating agency, for as recently as October 1973 he had expressed dissatisfaction with the delays in aid disbursement and had demanded in Cabinet that yen loan agreements and payments progress be reported to him directly.[28] There may have been more immediate issues in his mind, since he was apparently keen to press ahead with economic cooperation with Brazil, a goal which the new agency could assist, and his pending visit to Southeast Asia in January 1974 demanded some kind of symbolic aid policy iniative.[29]

Tanaka's decision in late December to institute the new ministerial post was unexpected, although the idea itself had already been tentatively raised by Minato. The MFA had always opposed the suggestion on the grounds that, unless the minister was a junior or assistant minister, the position of the Foreign Minister could be compromised. It is possible that the success of Miki Takeo's trip to the Middle East between 10-28 December 1973 as Special Ambassador encouraged Tanaka, for some of his statements about the Foreign Minister's load indicated that he thought the burdens of that office should be lightened. Legislation which lent force to Tanaka's wishes was already before the Diet. Presented in February 1973 but still pending in December it provided for a system of Cabinet Councillors (*naikaku sanyo*) to assist in high-level policy, including economic cooperation. They could have acted as advisers to the new minister.[30]

The proposal for a minister was obviously unworkable. Not even his responsibilities were clearly delineated and the *Nikkei* of 29 December described the position as 'pathetic'. The lack of any detailed preparation suggested that Tanaka used the plan with some success to force MFA agreement on the agency question, while Fukuda and his fellow ministers took the opportunity to undercut Ōhira, who was then regarded as the likely successor to Tanaka. Tanaka was also able to publicise both initiatives during his difficult trip to Southeast Asia in January 1974. The bill to create a ministerial post was quietly shelved: it was presented to the Diet separately from that providing for a new minister for national development (contrary to the earlier ideas of the Cabinet Secretariat[31]), but was never debated.

Drafting the JICA Legislation: Conflict and Compromise

The compromise between bureaucratic and political interests presaged an uncertain future for the new agency. Guidelines for its tasks, for the pattern of control over it between ministries and for its relations with other aid-financing bodies, were only outlines, but officials acted swiftly to prepare legislation to be introduced at the forthcoming 72nd Diet session. A small team was set up for this purpose in the Foreign Minister's Secretariat, headed by Yanagi Kenichi, Director of the Economic Cooperation Bureau's First Technical Cooperation Division. It began work on 4 January 1974 on the first draft of the proposed bill — not a very arduous job, but important in allowing the MFA to put its early imprint on the shape of the agency. The LDP Special Committee on Overseas Economic Cooperation, assuring its role as 'promoter' of the new moves to unify aid policy, quickly associated itself with the plans, announcing on 30 December that it would push ahead with drafting a new basic policy for aid.

Subsequently, a senior drafting committee was formed in the Councillors' Office of the Cabinet Secretariat with a strictly limited membership. Because of the controversy surrounding the legislation, members were restricted, by Prime Ministerial directive, to the relevant bureau directors (or their equivalent) in the responsible ministries. The chairmanship of this group fell to the Director of the AMA's Administrative Management Bureau, Hirai Michirō.

This appointment was rather unusual, but significant in having drafting problems resolved quickly. Legislative work in the Councillors' Office usually came under the jurisdiction of the bureaucratic appointee to the Deputy Directorship of the Secretariat (there was also a political appointee), but Hirai was well acquainted with the problem in his capacity as the officer in charge of reorganising government administration. While perhaps the Prime Minister's own choice, he was a bureaucratically neutral chairman. Nevertheless, his status as an AMA official meant that his authority over the committee was legally tenuous; he had to 'borrow' the authority of the Deputy Director and rely primarily on persuasion to effect changes in the draft of the bill.

The Hirai Committee was small, very powerful and designed for quick and effective policy work. This was important since time was at a premium, and many of the problems associated with JICA in 1976 were a consequence of truncated committee deliberations. The committee met a few times per week over January and the first half of February 1974, a period of intricate legislative work and continued bargaining, conducted both formally within the committee and informally between desk officers in the ministries. The lines of argument were explicit from the outset. The problem was how to reconcile the short-term objectives of MITI and the MAF with the MFA's longerterm conceptions of Japanese economic cooperation. As it turned out, the serious disagreements were not about relations between Japan and the developing countries but were determined largely by domestic factors, as some observers had rightly predicted.[32] The time available for drafting was extremely short and the protagonists were set in their aims. Debate was heated and the unsatisfactory compromises eventually reached in the JICA bill were testimony to the confrontation between entrenched ministry positions.

Several issues stood out, namely the scope of agency business, administrative control and the relations between the agency, the ministries, other agencies and private enterprise. The broad structure of the agency had been decided at ministerial level, and technical cooperation managed by the former OTCA was to be incorporated in the work of a number of departments, as were the functions of the JEMIS. The Japan Overseas Cooperation Volunteers was retained as a separate organisation as it had been under the OTCA. Those JODC operations relating to agriculture and mining and industry were included, and new departments were to be set up for them.

Dispute between ministries arose especially in regard to Article 21 of the bill dealing with 'Scope of Business', particularly paragraph 1.3, which detailed the 'new functions' of the agency — the provision of loans for overseas development projects. There was no real difficulty over paragraphs 1.1, 1.2 and 1.4 of Article 21, which concerned general technical cooperation, despatch of technical experts and acceptance of trainees, and emigration respectively. The new loans function was the direct result of original MAF and MITI plans for undertaking overseas resource projects, but all ministries tried to use the article to expand or defend their own jurisdiction in aid affairs. The final wording of the article was confusing, imprecise and severely limiting in scope.

It was decided that the new agency would provide loans only when the Eximbank or the OECF made loans to or investments in the project in question and when finance could not be obtained from those bodies for some components of the total project. These conditions were designed to protect the interests of the bank and the OECF and to prevent future JICA work restricting their funding. JICA was limited specifically to financing 'related facilities . . . concomitant to development projects' and 'experimental projects', a provision which has since caused misunderstanding about which projects can be supported, a consequent scarcity of projects and slow disbursement of available JICA loan funds. JICA had to wait for OECF or Eximbank participation (except in the case of 'experimental projects'), but could still only finance the fringes of a project, the 'related infrastructure', such as roads and bridges associated with large construction sites.

The committee also opened the way for the agency to undertake feasibility studies by agreeing that it could carry out surveys and technical guidance necessary for JICA projects, but one of the main controversies concerned the sectors which these 'related infrastructure' and 'experimental projects' would include. Agriculture, forestry, mining and industry were the first and most obvious areas agreed upon, but the final version of the JICA Law also included the objective of cooperation in 'social development', that is 'culture, transportation, communication, health, living environments, useful for the promotion of the welfare of the inhabitants in developing areas' (Article 21.1.3 (a)). This, as we shall see, was the outcome of pressure from ministries not included in the committee.

Another dispute within the Hirai Committee related to supervision of the agency and to consultation between ministers over jointly controlled functions. Each ministry was determined to exact maximum benefit from the discussions and, once again, the resulting compromise was a classic example of bureaucratic competition taking precedence over commonsense. The chairman had to take his efforts at coordination to vice-ministerial level before agreement could be reached.

After eighteen months of competition over this issue, there were strong differences between the MAF and MITI about the extent of their existing authority and how this impinged on the agency's new financing provisions. MITI insisted that trade extended into the agricultural development aspects of JICA's work, but the MAF countered by claiming that it was the relationship of development to food policy rather than to import policy which was pertinent. The MFA was keen to extend its powers of supervision into all areas of the agency's work. This was evident in its desire to control development surveys, although the MAF and MITI opposed this. The committee realised that surveys carried out on a project basis would overlap the boundaries of all three ministries, and they were therefore made a 'joint jurisdiction' (*kyōkan*). By 1979 there was a tangled series of procedures for budget approval and for implementation of surveys.

The Law (Articles 42-3) stipulated that the Minister for Foreign Affairs had to consult the Minister of Finance on all financial and accounting matters and the Ministers of International Trade and Industry and Agriculture and Forestry on all business connected with their respective functions. The latter two ministries had equal competence with the Minister for Foreign Affairs in respect of development financing. Even where the Foreign Minister was the only competent minister (as in ordinary technical cooperation), the system of mandatory consultation restricted him or his officials in making policy and set up a continuing cycle of claim and counter-claim in territorial disputes.

The committee also debated what shape the executive structure of the agency should be, for this would seriously affect the way ministries influenced its day-to-day administration. It was normal practice for ministries involved in an agency's work to see their interests represented to some extent by the appointment of their own officials (or former officials) to executive positions. This was part of a widespread practice in government of appointing retiring senior bureaucrats to statutory corporations, one aspect of *amakudari*, or 'descent from heaven'. It had its problems, of course, but was useful to the agency, since men with experience of ministry affairs entered top management jobs.[33] At first a chairman-directors (*kaichō-riji*) format, similar to many existing statutory corporations, was considered, but it made a balance of interests difficult to achieve. The chairman would represent the controlling ministry (the MFA), while MITI and the MAF would be given only directorships. To place the latter two on an equal and elevated footing, the alternative of having a president and two vice-presidents emerged as the best course and was incorporated in Article 8 of the bill. Settling on the number of directors, however, proved a more complicated problem. The large number of ministries and agencies, including those being abolished, which demanded a place meant that instead of the three or four directorships normally allocated to a statutory body, twelve were assigned to JICA.

Once these issues had been dealt with by the committee, the LDP's PARC Deliberation Commission considered the bill for approval before it went to Cabinet. This was the usual procedure for government policy, allowing the airing of pressure group and factional attitudes, and eventually achieving broader acceptance for legislation within the governing party. It also allowed a chance for direct pressure, and other ministries and their supporters in the party took the opportunity to force changes in the draft financing functions of the agency. There seems to have been a delay of a week in the timetable envisaged for the bill in late January. The *Nikkei* on 27 January reported that the bill was drafted and would be approved by Cabinet on 8 February. While there was no mention of the 'social development' aspects of JICA's work at the earlier date, by 7 February they had been included.

Ministries with small economic cooperation programmes, excluded from the Hirai Committee — the Construction, Transport, Posts and Telecommunications and Health and Welfare Ministries and the Science and Technology Agency — wanted to participate in the work of the agency but were forced to wait until this late stage in drafting to put their view. They especially wanted to extend the scope of agency financing and to clarify the meaning of 'infrastructure' financing and the relation between it and loans to actual industries or projects. The LDP Deliberation Commission advanced the case, on behalf of these ministries, that 'international cooperation' meant much more than assistance to infrastructure projects in agriculture, mining and industry. Consequently, the concept of 'social development' was introduced to the body of the draft, in Articles 1 and 21.1.3. There was insufficient time to resolve doubts about precise interpretation of the concept, or to make substantial changes to the wording of the bill (such as in articles referring to jurisdictions), so JICA's confused functions stemmed partly from this late interference in decision-making already rushed and tense.

The relationship between agency financing and investment by private enterprise was also raised. The Hirai Committee had determined that, in principle, the agency would not fund projects where private firms were in a position to profit, and that projects undertaken on a largely private basis should be financed by other institutions, such as the Eximbank. JICA was to support projects, or parts of projects, with an extensive public or 'social capital' content. This was neither entirely what the MAF had originally wished for, nor what Tanaka had envisaged.

In respect of private firms, however, the Law remained equivocal. The OECF and the EPA argued vigorously to prevent JICA financing foreign governments directly, so that the OECF would not be cut out of implementing future government loans, but no clause was placed in the bill to define who should receive agency funds. It was left to inter-ministry memoranda on JICA 'financing guidelines' to stipulate that only Japanese companies could receive JICA loans. Therefore, the aim of the original drafting committee to avoid links between JICA and private enterprise was ignored and the agency was restricted instead to offering finance to Japanese companies or to persons engaged in development projects overseas. Furthermore, no criteria were established to measure the public benefit of such projects. JICA operations were thus severely curtailed and made subject to the whims of external decision-makers: it could finance only 'related facilities' or 'experimental projects' (and needed associated OECF or Eximbank funding in the former) and had to depend on investment decisions by private enterprise before it could become involved in either.

In all, MAF, MITI and MFA conceptions of economic cooperation were not satisfied in JICA, and the later effectiveness of JICA as a financing agency rested primarily on the initiative of private businesses. The scope of JICA operations was still unclear in 1979 and arguments between ministries over jurisdictions prevented the smooth implementation of agency programmes, as will be pointed out in detail in Chapter 4. While initially political pressures forced a compromise in the budget context, late political interference in debate on the new agency's structure and responsibilities only complicated the agency's tasks and left several unresolved boundary issues.

The Diet Debates

After five weeks of committee bargaining, Cabinet ratified the JICA bill on 15 February 1974. It was presented to the Diet on 18 February, passed by the House of Representatives on 14 May after deliberations in the Foreign Affairs Committee (where most of the debate took place), and by the House of Councillors on 27 May. It was promulgated as Law No. 62 on 31 May 1974.

By the time the bill reached the Diet, basic organisational questions had been settled. The capital of the agency was set by Article 4 at 4 billion yen and, with the funds of the dismantled agencies (the OTCA, the JEMIS and part of the JODC) added, initial capital totalled 22.4 billion yen. The budget for 1974, consisting of operational expenses and capital transferred from the 1974 budgets of the absorbed agencies plus a new budget for the agency, amounted to 27.37 billion yen. Staff were to number 900-1,000, including 500 from OTCA, 420 from the JEMIS and about 100 to be recruited from ministries to work in the new financing and survey departments. A decision on the English name for the agency was not made until July.

The pattern of parliamentary support was predictable. The Japan Socialist Party (JSP), Japan Communist Party (JCP) and Kōmeitō all opposed the JICA bill, although their opposition never endangered its passage through the Diet. Their specific objections were also expected: criticism of the overseas expansion of Japanese capital as represented by JICA and the neo-colonialist nature of resources development projects. The Kōmeitō differed slightly with its emphasis on human rights, which, it claimed, Japan's economic cooperation failed to protect. The Democratic Socialist Party (DSP) supported the bill.

The debate on the bill turned out to be one of the most comprehensive on foreign aid held by the Japanese Diet, but it excited few passions. Some issues — the relationship between the new agency and the bill for an economic cooperation minister, the control of the agency, resources development and development import, JICA and private firms, the administration of Japanese aid as a whole — were discussed at length, and sometimes with vehemence. Much of the criticism of the agency and of the proposed minister (the bill for which was also before the Diet), was levelled at the increased powers which could pass to the Prime Minister and arose from doubts about Tanaka's personal motives. JSP members Wada Sadao and Dōi Takako both adopted this line in their speeches on 4 May and 10 May 1974, in the House of Representatives and its Foreign Affairs Committee respectively. An

LDP member, Ishii Hajime, alleged that the agency was supervised by too many ministries and that it would quickly lose its administrative efficiency. He considered that there were too many areas of joint control, too many directors and too many officials transferred from the ministries.[34] One Kōmeitō member, Watanabe Ichirō, in a lively Foreign Affairs Committee speech on 8 April, called JICA a 'freak'. Other charges that employment conditions varied between sections of the staff were pertinent, for there had not been enough time for the drafting committee to sort out the details of combined ministerial control of the agency, let alone inconsistent salaries and conditions between former OTCA and JEMIS staff. There was, in fact, some irony in the failure of the Hirai Committee to settle staff questions, while creating an agency destined to serve as a training ground or post-retirement employment for officials of its many supervising ministries. The OTCA had a history of staff activism which had culminated in a strike in 1970 over the prevalence of precisely this problem, 'descent from heaven' appointments.[35]

The most extensive part of the parliamentary debate was on the concept of 'development import' and on the degree of JICA involvement. The discussion produced some hasty justification by government witnesses for those JICA functions which resembled development import, and revealed interesting differences between the MFA and the MAF over how far development import should be pursued and about the relation of development import to domestic food policy. They were differences which had existed since the MAF first requested an agency in 1972 and were evident in the MAF claim to JICA as an adjunct to the government's agricultural and food policy which dealt primarily with increasing Japanese self-sufficiency. The development aspect of the agency's work was of secondary importance to the domestic concerns of the MAF. In contrast, it was of high priority for the MFA, even though there was no objection to development import accompanying bilateral economic development aid. From the testimony of MFA officials, however, it became clear the JICA would not be restricted to work in developing countries but that a place like Australia was, among others, a likely target for development finance. This admission had been foreshadowed in Prime Minister Tanaka's NHK broadcast of 27 December; it tended to support opposition claims that JICA was created above all to develop resources for Japan's benefit, and detracted from the MFA emphasis on the development impact of the new agency.[36] The vague wording of Article 1, which specifies that JICA can finance projects in 'developing and other areas' drew criticism that

JICA would not be interested in assisting only the less developed nations.

While it was lengthy, the debate did not reveal any familiarity of members with aid or with development arguments. This pattern was not unusual for Diet debates, not normally renowned for their contribution to informed discussion. It was difficult, if not impossible, for the opposition to make more than token criticisms. Complaints about Japanese aid performance and aid administration reiterated well-worn arguments about poor performance, lack of coordination and non-existent policy guidelines. Some censure arose of the fact that loans were to be made to Japanese companies, but it was not driven home with any conviction; perhaps members were aware of the indirect political benefits of that rule. Answers by government witnesses were bland and repetitive, composed from a comprehensive set of questions and responses prepared by the MFA Economic Cooperation Bureau. There was no extended debate on control of agency business by the ministries or on the relationship between JICA and the OECF or the Eximbank. Questions by the opposition parties were not penetrating and tended to be set-piece performances aimed at scoring parliamentary points rather than a concerted attempt to take the government to task over its aid policies. The debate was dull and inconclusive and failed completely to solve any of the glaring problems within the JICA bill. The Diet was ineffective in the formation of JICA, and the case well supports Baerwald's contention that 'most committee work in the Diet is an exercise in futility'.[37]

The JICA bill brought out more than anything else the parochialism of the party system on aid questions. Because its passage through the House of Councillors ran very close to the suspension of the Diet session, some hard bargaining was necessary to force the bill through on time, according to Sakurauchi Yoshio, a former Minister of Agriculture and strong supporter of the MAF scheme. One interesting result of the debates was a 'supplementary resolution to the JICA bill', supported by all parties except the JCP. It was an unabashed concession to the desire by opposition parties and some elements of the LDP to calm domestic opposition to JICA, and clearly expressed *pro forma* compromises on the role of private enterprise in aid, JICA relations with other agencies, MFA relations with other ministries, aid and Japanese agricultural policy, and inconsistencies in staff conditions. It enshrined the truth that consensus-building needs time, a commodity of which the creators of JICA had had very little.

Choosing Personnel

The choice of candidates for executive positions was settled unofficially in May, after the bill had been passed by the Diet. The *Nikkei* of 30 April commented that the MFA, MITI and the MAF were competing strongly for the available executive posts. It seemed that one directorship and one auditorship (*kanji*) would go to the representatives of the MOF, but that the MFA was working hardest to secure positions in the agency. It needed to find jobs for the executives of the two organisations under its control which had been abolished (the OTCA and the JEMIS) and in any case the MFA traditionally had few posts available to offer its retiring senior officials. Naturally it wanted to increase these opportunities.

These were plausible reasons. The MFA did not have the scope which home-based ministries, such as the MOF, MITI or the MAF, had for placing retiring officers in semi-government or private organisations. Some had argued that the MFA saw JICA from the outset as a rich pasture for its 'old boys', but as it transpired the balance achieved between ministry appointees to JICA directorships was a very fine one and the MFA gained no particular advantage.

It was never certain during the drafting of the bill that the position of president would go to an official, even one from the MFA. Some politicians and businessmen wanted a non-government candidate, as Nakayama Soppei, the former President of the OTCA, had been. It was Nakayama himself, however, who opposed this idea, by stressing that more than a figurehead was required. Officials agreed that a working president was essential to the success of JICA and that an experienced bureaucrat would be the best qualified. At this point, it appeared probable that the top post would go to an MFA choice.

Three men, Hōgen Shinsaku, Asakai Kōichirō and Mori Haruki, seemed the most likely contenders, according to the *Nikkei* of 30 April. Asakai and Mori were both well versed in aid. The former was an adviser to the Foreign Minister and a Director of the Bank of Tokyo, while Mori was Ambassador to Great Britain with extensive experience in economic cooperation, including a period as Japan's first Ambassador to the OECD. Of the three, Hōgen had the least experience in economic cooperation and aid,[38] but he had been forced to resign from the vice-ministership only a few months before the JICA presidency became an issue. Before he left office, he had discussed possible candidates for the JICA position with Mikanagi Kiyohisa, Director of the Economic Cooperation Bureau, and with Katori Yasue, Director of the

Minister's Secretariat, but at that stage was not in the running himself. In April, however, Katori suggested that the recently dismissed Hōgen be offered the job and both Ōhiro and Tanaka agreed to this by mid-May.

JICA executives were a mixed group, Hōgen became President while the Vice-Presidents were Hisamune Takashi, Director of the Overseas Fisheries Cooperation Fund and a former Director of the MAF's Economic Affairs Bureau, and Inoue Takeshi, director of an engineering firm and a former Director of MITI's Economic Cooperation Department. This balance at the top of the executive structure was maintained throughout. The demands of the abolished agencies were met by the appointment of four former directors and auditors from the OTCA, four from the JEMIS and one from the JODC. Of the total of eleven directors and three auditors, six were former MFA officials or belonged to agencies under its control, one previously worked in the MOF, two in the MAF, two in MITI, one in the Ministry of Construction and one in the Ministry of Posts and Telecommunications, while one was a former member of the House of Councillors.[39]

Staffing of the agency was straightforward, but also reflected care in weighing, and in satisfying, ministry interests. OTCA, JEMIS and JODC staff were moved across and others were transferred from the MAF and from MITI. Different departments of the new agency came under the influence of their responsible ministries through personnel transfers and their resultant effect on policy. While Emigration, Training and Medical Cooperation Departments took in staff from the JEMIS and the OTCA, the new departments had new personnel. The General Affairs Department was headed by a former official of the MFA, Personnel Department by an OTCA officer, Mining and Industry Departments by men from MITI, Agriculture Departments by officials from the MAF, and the Social Development Cooperation Department by a former officer of the Ministry of Transport. The general overview of JICA departments exercised by ministries was reinforced by the way in which directors' responsibilities were allocated. Thus MOF influence in the Accounting Division was said to be strong, especially since it reported to a director who was a senior official of the International Finance Bureau before his appointment to JICA.

The New Agency

Public appraisal of the agency was predictably generous, but the problems thrown up by JICA's troubled birth could not be, and were not, ignored. The new Foreign Minister, Kimura Toshio, saw JICA making up for the 'insufficiencies of the implementing machinery' and providing links between technical and capital cooperation,[40] but Nakayama Soppei, immediate past President of the OTCA, was not as optimistic:

> The merging of a number of aid agencies into one has no signifi-
> cance unless it results in better and more effective development
> assistance. I encountered difficulties arising from the Government's
> vertical division of responsibilities even when I was managing only
> one agency, the OTCA. Now that the new Agency's scope of activi-
> ties had been broadened much beyond that of the OTCA, I think the
> difficulties are likely to be even more acute.[41]

The new President, Hōgen Shinsaku, consistently expressed his desire to use the agency to pursue the economic development and well-being of the developing countries. At the same time, he appreciated the restrictions arising not only from within a new body but also from the environment of the aid bureaucracy as a whole. 'It is necessary,' he said, 'to build unity within the new Agency as quickly as possible, but it is silly to think that this can be done immediately . . . we must be patient.'[42]

Other commentators were less considerate. MAF officials remarked, both in print and in private, that the agency once in operation was not what they had expected, because JICA's capacity to harness technical cooperation to development financing was not effectively exploited. In terms of bureaucratic influence, however, the MAF benefited from JICA's creation, since its share in the agency's administration brought the MAF closer to the centre of Japan's aid bureaucracy. Although this did not lead to participation in the making of government loans policy, it was nevertheless a foothold in what was to become a far more active side of Japan's aid programme. The MFA also fared well, by securing majority control of JICA, even though it had not set out to do so and had originally opposed the very notion of another government aid organisation. MITI, however, lost one of its subsidiary agencies (the JODC) and won little more influence over JICA than it had had over OCTA. It had to be content with new development financing facilities.

It can take several years for a newly independent agency to find its feet and settle into an established relationship with surrounding mini-stries and its fellow executive agencies. JICA was not without such difficulties, for it was created at a time of deepening recession in Japan following the oil shock and a dramatic decline in overseas private investment from $3,647.5 million in 1973 to $1,038.5 million in 1974 and $352.4 million in 1975 (see Table 1.1). JICA development financing was, as explained above, dependent on the initiative of private investors, and reduced investment meant that JICA lending ran at a very low ebb. This led to a rapid cut in JICA's development finance budget from 7,200 million yen in 1976 to 5,000 million yen in 1977 and 500 million yen in 1978.

There were also problems in defining JICA's authority in the area of development surveys and financing, stemming mainly from the vagueness of the JICA bill but also from the lack of direction in the largely uncharted administrative waters lying between JICA and the OECF. The entire future of the JICA role in development financing, and its relations with the OECF and the Eximbank, depended on the meaning of 'related facilities' and 'experimental projects' in Article 21 of the JICA Law.

JICA and Policy-making

The JICA study is instructive in several ways. It tells us how easily policy objectives, particularly those in an area peripheral to the main interests of powerful ministries, are diverted by temporary pressures and political whim. It was a good example of the subordination of aid policy to predominant ministry concerns, even in the original MAF and MITI proposals, which were directed mainly at solving donor policy problems. That aid policy later became the active focus of the argument was due largely to the actions of two politicians: the initiatives by Minato and Tanaka's use of aid for his own political and diplomatic ends. The ministry with the obvious rationale for linking the agency question with ongoing aid programmes — the MFA — did so only by default, and too late to greatly affect the final decision. The JICA issue was really a domestic problem, concerning bureaucratic power and the domestic effects of external policies.

The decision-making process itself was important in determining the shape of the agency. The constraints of the normal budgetary process prevented an early appraisal of the merits of ministry proposals. Political pressure broke the impasse, and conflict was constrained later because participation in the political decision and legislative drafting was restricted. Compromise became easier when the options were limited. Only a few politicians were actively interested and LDP factions were not openly involved, except informally in the coalition of Fukuda, Hori and Kuraishi.[43] The relevance of high-level committees to abnormal policy situations was clear.[44] When participation widened, however, and the LDP and other ministries interfered, priorities became confused and the draft legislation suffered as a result.

Timing had a marked effect on the outcome. Two years of preparation by the MAF ended in a political decision which took one week and in drafting which lasted one month, when completely new options were raised and adopted without detailed consideration. The deadline for the budget decision and the presentation of the bill to the Diet left no opportunity for ironing out the conflicting interpretations of the functions of the agency. This was linked to the fact that perceptions of the policies and the issues in the whole JICA story never met. Decisions were made on temporary compromises between completely divergent sets of principles, bureaucratic interests and political purposes.

Notes

1. John White, *The Politics of Foreign Aid* (London, Bodley Head, 1974), p. 254.

2. Anthony Downs, *Inside Bureaucracy* (Boston, Little, Brown and Company, 1967), pp. 220-2.

3. Two contrasting works which develop this theme are Michel Crozier, *The Bureaucratic Phenomenon* (Chicago, University of Chicago Press, 1964) and Peter Self, *Administrative Theories and Politics* (London, Allen and Unwin, 1972).

4. Herbert Simon, 'Birth of an Organization: The Economic Cooperation Administration', *Public Administration Review*, vol. XIII (Autumn 1953), pp. 227-36.

5. Much of my information on the MAF is based on Ashikaga Tomomi, *et al.*, 'Kokusai kyōryoku jigyōdan' ('The Japan International Cooperation Agency'), three round-table discussions which appeared in *Yunyū shokuryō kyōgikaihō* (April-June 1976), between officials of the MAF closely involved with the JICA establishment. I shall refer to it hereafter as *Bulletin*. Other information was derived from interviews with officials of the MAF, MFA, MITI and the EPA in 1976-7, but references to these are not all made specifically in the notes.

6. E.A. Saxon, *Japan's Food Gap and Trade in Farm Products* (Canberra, Bureau of Agricultural Economics, Occasional Paper No. 42, 1977) has food import figures, and Tsūshō sangyōshō, *Keizai kyōryoku no genjō to mondaiten* (Ministry of International Trade and Industry, *Economic Cooperation: present situation and problems*, hereafter MITI, *Economic Cooperation*) (1974), contains data on aid by sector.

7. *Bulletin*, April 1976, pp. 1-11.

8. John Creighton Campbell describes the research grant as a useful tactic for ministries to get their foot in the budgeting door. See his *Contemporary Japanese Budget Politics* (Berkeley, University of California Press, 1977).

9. Fred H. Sanderson, *Japan's Food Prospects and Policies* (Washington, DC, Brookings Institution, 1978).

10. The term *'kaihatsu yunyū'* ('development import') as a means of providing stable resource supplies was first mentioned in the 1960 MITI, *Economic Cooperation*, p. 50. It dropped out of usage in the mid-1960s because of its exploitative stigma, although the term reappeared in the 1967 edition. Development import of agricultural products was put into practice in such projects as the Mitsugoro maize farms in southern Sumatra, a joint venture between Mitsui Bussan and the Indonesian Kosgoro group, with OECF funding. One study is Alan G. Rix, 'The Mitsugoro Project: Japanese Aid Policy and Indonesia', *Pacific Affairs*, vol. 52, no. 1 (Spring 1979).

11. E.A. Saxon, *Japanese Long-Term Projections Relating to Food and Agriculture* (Canberra, Bureau of Agricultural Economics, Occasional Paper No. 38, 1976).

12. *Asahi shimbun* (hereafter *Asahi*), 31 July 1973 and *Bulletin*, April 1976, p. 14.

13. Details in Minato Tetsurō, *Nihon no ikiru michi: kokusai kyōryoku no kihon seisaku o mezashite (Japan's lifeline: towards basic policy for international cooperation)* (Tokyo, Kokusai kyōryoku kenkyūkai, 1975), pp. 99-100.

14. MFA attitudes were based mainly on a policy statement reported in *Nihon keizai shimbun*, 8 July 1973, statements by Mikanagi Kiyohisa (Director of the MFA Economic Cooperation Bureau) in the House of Representatives Foreign Affairs Committee on 5 April 1974, and interviews with several MFA economic cooperation officials (past and present) on 14 June, 1 July, 11 and 13 August 1976.

15. *Bulletin*, May 1976, pp. 1-5. The Animal Industry Bureau and Forestry Agency were very much concerned with promoting and protecting domestic producers and were therefore suspicious of the development of overseas import sources.

16. The portfolios of Foreign Affairs and International Trade and Industry remained with Ōhira Masayoshi and Nakasone Yasuhiro respectively. Kuraishi himself, in an interview reported in Minato, *Nihon no ikiru michi*, p. 173, agreed that these changes were decisive.

17. Campbell, *Contemporary Japanese Budget Politics*, p. 260.

18. See interview with Kuraishi in Minato, *Nihon no ikiru michi*, pp. 172-3, and *Bulletin*, May 1976, p. 5. The text is in Minato, pp. 104-6.

19. Interview, 23 December 1976.

20. These paragraphs are based on an interview on 23 December 1976 with one of those present at the meeting.

21. *Asahi*, 27 December 1973. The *Yomiuri shimbun* (hereafter *Yomiuri*) of 28 December stated that Fukuda wanted to set up agencies only if 'those that have run their course change their name'. He was referring specifically to the Agricultural Machinery Corporation but may also have been hinting at OTCA and JEMIS.

22. *Yomiuri*, 27 December (evening), and 28 December 1973.

23. Tanaka retorted by accusing all MFA officials of being useless and acting 'like a lot of self-important feudal lords' (*Yomiuri*, 28 December 1973).

24. *Nikkei* and *Yomiuri*, 29 December 1973, and Minato, *Nihon no ikiru michi*, pp. 107-8. The *Gokajō no goseimon* referred to a very broad set of guidelines providing for what would necessarily be an extremely complex task.

25. The ability of the MOF to produce this sum for the agency at the very end of the budget negotiation process surprised many people. The then Finance Officer of one ministry admitted that he did not know where the money had come from and was sure it was not from funds already allocated to the MFA

technical aid budget, even though details of that were not finalised until some days after the Budget was announced. On the MOF use of reserve funds, see Campbell, *Contemporary Japanese Budget Politics*, Ch. 7.

26. The report, released in June 1974 and which recommended a large number of changes to technical cooperation management procedures, was rendered largely useless by JICA's establishment.

27. *Nikkei* and *Yomiuri*, 28 December 1973. The *Yomiuri* called the plan Tanaka's *'medama shōhin'* or 'prize merchandise'.

28. *Asahi*, 31 October 1973.

29. Suggested in *Nikkei*, 28 December 1973, and in interviews with MOF and MFA officials and a former AMA official on 6 August, 11 August and 16 November 1976.

30. The *sanyo* idea had its roots in the 1966 report of Rinji gyōsei chōsakai (Special Commission on Administration) on the functioning of Cabinet, in which a system of Cabinet assistants was recommended. The Administrative Management Committee (Gyōsei kanri iinkai), a high-level committee set up in July 1965 to report on administrative management and reorganisation to the Director of the AMA, also decided that the Cabinet Secretariat was overloaded and, in a report presented on 27 December 1972, suggested that Cabinet Councillors be appointed to help coordinate special policy areas such as environment, economic cooperation and budgeting.

31. The Secretariat originally suggested one bill increasing the number of ministers from 19 to 21, to include a new Minister for National Development and Minister for Economic Cooperation.

32. See the editorial in *Kokusai kaihatsu jānaru*, 25 January 1974, p. 2. A letter in the same issue (p. 36) from an anonymous university professor pointed clearly to the problems of reconciliation. He declared that ministries would act only in their own interests and that the political compromise that gave rise to the agency would only lead to an emphasis on Japan's interests. He also foresaw JICA as a dumping ground for older MFA officials.

33. Chalmers Johnson, *Japan's Public Policy Companies* (Washington, DC, AEI-Hoover, 1978), Ch. 5.

34. *Gaimu iinkaigiroku 17-gō*, 5 April 1974, pp. 7 and 11. See also Kawakami Tamio (JSP) in *Gaimu iinkaigiroku 18-gō*, 8 April 1974, p. 1.

35. On the strikes, see debates in the Shakai rōdō iinkai (Social-Labour Committee) of the House of Representatives, 63rd Diet Session, June 1970, and the House of Councillors' Kessan iinkai (Audit Committee), November 1970.

36. Accusations by Nagamatsu Eiichi (DSP), *Gaimu iinkaigiroku 24-gō*, 10 May 1974, pp. 1-3.

37. Hans H. Baerwald, 'Committees in the Japanese Diet', unpublished paper, n.d.

38. Hōgen Shinsaku was born in 1910 and graduated from Tokyo University's Law Faculty in 1933. He joined MFA and made a career in European and American affairs, being Councillor in the Japanese Embassy in Moscow, Ambassador to Austria, and the Ambassador to India for one year in 1968. He became Vice-Minister for Foreign Affairs in April 1972 but was dismissed in February 1974 by Foreign Minister Ōhira. The incident arose originally from a mistake on the part of the then Ambassador to Washington, Yasukawa Takeshi, who stated incorrectly during a visit by Ōhira that the 1973 Nixon-Tanaka communiqué included words to the effect that the Emperor would visit the US 'during 1974', a statement which Ōhira accepted and repeated. This brought a formal reprimand from the Prime Minister to the Ambassador and pressure on Ōhira to make changes in the top echelons of the MFA, already under attack from a number of quarters for other mistakes, including a failure to notice inconsistencies in the

texts of the Japan-Soviet joint communiqué of October 1973 (see *Asahi*, 25 October 1973). On 18 February 1974, Ōhira called in Hōgen and asked him to step down. He was made an adviser to the ministry and given an office in the MFA building. Nagano Nobutoshi describes the events in *Gaimushō kenkyū (A study of the Foreign Ministry)* (Tokyo, Simul, 1975), pp. 28-34.

39. The directors chosen were: Mikanagi Kiyohisa (Director of the MFA Economic Cooperation Bureau), Itō Takuya (Director of JEMIS and former MFA official), Kondō Michio (Director of OTCA and former MOF official), Yoshihara Heijirō (Director of OTCA and former MAF official), Endō Kanji (MAF Councillor), Nakanishi Kōichi (Director of OTCA and former MITI official), Nagao Mitsuru (Director of Civil Engineering Research Institute and former official of the Ministry of Construction), Saitō Minoru (Director of JEMIS), Toyama Motohiko (Director of JODC), Shinshi Masao (auditor of JEMIS), Hitomi Yoshio (official of Telegraph and Telephone Corporation). Auditors were Yamamoto Toshihogi (LDP Member of the House of Councillors), Okada Katsugi (auditor of JEMIS) and Moriya Eitarō (auditor of OTCA).

40. 'Inauguration of the Japan International Cooperation Agency', *Look Japan*, 10 August 1974, p. 3.

41. Tanaka Kakuei and Nakayama Soppei, 'New Vigor and Direction in Japanese Development Assistance', *Look Japan*, 19 August 1974, pp. 1-2 at p. 2.

42. *Kokusai kyōryoku*, no. 239 (August 1974), p. 8.

43. Minato Tetsurō is said by one member of the same Nakasone faction (interview, 23 December 1976) not to have consulted either Nakasone or other faction leaders about his role in the decision. This was to avoid delays which would have resulted from such discussions.

44. See the case analysis by Fukui Haruhiro in his 'Tanaka Goes to Peking: A Case Study in Foreign Policymaking' in T.J. Pempel (ed.), *Policymaking in Contemporary Japan* (Ithaca, Cornell University Press, 1977), pp. 60-102.

Part II:
The Domestic Politics of Foreign Aid

3 AID AND THE GOVERNMENTAL PROCESS

How did policy evolve from this tangled system? Where did aid fit in government? John White, in his study of the politics of foreign aid, concluded that the makings of an aid policy lie within 'the rather narrow institutional environment' in which aid institutions operate, that people, rather than agencies or governments, determine the forces which allow policy to evolve, in a continuous and cumulative process.[1] Beginning with a discussion of donor aid management, this chapter begins to probe this 'environment' as it was in Japan — initially in the bureaucratic setting — and examines the political context of the Japanese aid administration. Chapters 4 and 5 take the reader further into, first, the processes of the aid machinery and, secondly, aid budgeting.

Donor Aid Politics

While the aid official's immediate working environment may be narrow, the pressures on him are not. His is a classic dilemma: torn not only between the recipient's demands for aid on improved terms and the restrictions of available domestic resources, but also between the desire to serve the developmental objectives inherent in any aid policy and to accommodate aid with competing domestic policy interests. There is a constant tension about aid policy and a continual search for a new balance between aid and the other, often more pressing, duties of national government. This is an obvious issue for aid budgeting officials, but it touches others equally — the diplomat weighing aid requests, the trade officer assessing import controls, the planners fitting aid into their trade and economic models. Aid has always been seen both as a tool of policy and a policy in its own right, an approach which has created ambivalence about what aid is and does.

The Japanese aid system was centred on a small section of the national bureaucracy and drew in some of the most powerful ministries. It was also diverse and rather colourful in bureaucratic terms, embracing some unusual policy bedfellows. The structure of other donor administrations varied, from centralised systems with a single aid agency, to those displaying a highly dispersed array of controls (such as

83

Japan). The interplay of the three important distinguishing features of aid structures – (a) the location and exercise of political responsibility for aid, (b) the administrative form, and (c) the pattern of financial control – has been the dominant influence on aid programmes, although the origins of structures affected the way in which governments first approach the donor role. Strong ministerial leadership was the most effective although least consistent pattern, but only four donors (the United Kingdom, Germany, Sweden and the Netherlands) had a minister for aid. A number of others (such as Australia, New Zealand, Norway, Denmark, Canada and the United States) subsumed responsibility for aid under the portfolio of the Minister for Foreign Affairs. This made for a more elegant aid structure, but did not necessarily lead to smoother decision-making unless the agency had considerable influence over the continuity and use of funds, and the allocation of aid. Some donors, Japan included, had no defined political structure for aid, but relied on a dispersed administrative pattern to delineate political relationships.

Still, administrative machinery for managing aid policy usually rests uneasily in the donor's government structure. First, aid is a curious and, in many ways, unique policy area. While the indirect benefits to some segments of society (such as exporters) may be considerable, the primary reason for the existence of an aid policy lies outside the donor country and the direct beneficiaries cannot normally participate in the policy process. Nevertheless, although aid ostensibly caters first to an overseas clientele, domestic interests are well served and channels representing them are usually well defined. As later chapters will show, foreign aid is a prime example of a policy significantly affected by domestic structures and drawing energy from domestic forces.

Secondly, administrative structures for aid policy-making and implementation are never wholly independent. Their place in the national bureaucracy is not a strong one for their own *raison d'être* is external to the system and their domestic power is set within constraints determined by other relevant ministries and agencies. They have no power base of their own and the more dispersed the aid administration, the more acutely obvious this becomes. In addition, an aid agency, as John White puts it, 'has no natural allies, for whom the agency's activities are of vital concern, within the national political frameworks of the donor countries',[2] although this may be less obvious when aid is managed by several ministries, as in Japan. The allies an agency may attract at any time are not normally predisposed to combine or co-ordinate on behalf of the agency or its objectives. For example, despite the concentration of economic and technical resources, and energetic idealism, brought together by a separate ministry in Britain, the Overseas Development Ministry lacked a power base in Whitehall and, in the words of Judith Hart, a former minister, 'was an isolated ministry . . . Nobody cared very much what it did . . . It offended no one, and aroused neither enmities nor affection in Whitehall'.[3] Seers and Streeten, in their assessment of the Labour Government's record on aid, concluded that 'setting up an independent Ministry is merely an empty gesture unless there is real support within the Government for its policies'.[4] Judith Tendler found that the United States Agency for International Development (AID) was 'bullied' by other government agencies because of AID's very inability to muster domestic support. Lacking a domestic clientele, the agency invited 'incursions'.[5]

In Japan, in contrast to Great Britain, there was no conspicuous 'aid lobby', no widespread group of aid advocates among the government and the general public, a fact which some have attributed to the Confucian ethic of the Japanese people and their different attitude to *noblesse oblige*.[6] There are, of course, parallels between Japan's highly decentralised system and that based on a central aid agency, such as Tendler describes. Indeed, the nature of the foreign aid task, she concluded, encouraged decentralisation even in policy-making in an agency like AID. In Japan's case, however, the benefits of this 'organisational fit' (listed by Tendler as less formal procedures, constant contact with the recipient, a sense of organisation and of purpose) were not — as will be seen — as easily reaped.

Aid and the Ministry Environment

Aid policy was, of course, a responsibility of the Japanese Government and the hierarchies of parliamentary democracy in Japan, where executive power extended from the Prime Minister and his ministers to the ministries and their attached agencies, applied equally to aid as to other policies (see Figure 3.1). Official advice on aid was provided by an advisory council answerable to the Prime Minister and top-level coordination was left to Cabinet. In theory, policy implementing agencies were distinct from policy-making minstries but, in practice, the divisions were blurred. The relationships between Cabinet, the ministries and their agencies were formalised in a series of administrative laws and regulations, beginning with the National Constitution and extending to the Cabinet Law, National Administrative Structure Law and ministry

Figure 3.1: Japanese Aid Administration: Formal Outline

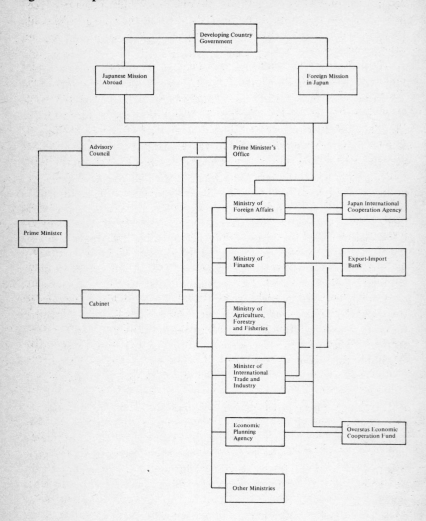

Source: Gaimushō, *Namboku mindai to kaihatsu enjo* (Tokyo, Kokusai kyōryoku suishin kyōkai), 1978, p. 706.

establishment laws. There existed no foreign aid legislation defining Japan's aid objectives or methods, although such bills were introduced in the Diet on occasion. Some officials felt that their own tasks could be appreciably assisted by being clearly defined in specific laws, but others were less sanguine.

The aid administration lay within the national bureaucracy, a system bound by tradition and popular myth, with a history of active social direction.[7] The elitest sentiments expressed by and about bureaucrats, even after the Occupation reforms had begun to democratise the powerful civil bureaucracy, exemplified the strength of convention. The popular literature would have it that a career in the MOF was the pinnacle of achievement, so much so that the image of the MOF in the popular mind was often equated with supremacy in policy-making.[8]

Ideas of or pretensions to elitism, however, did not of themselves bring power in government. In foreign aid, the MOF was one among many, although admittedly its control of budgets gave it principal leverage in interministry discussions. By way of contrast, the MFA's traditional ineffectiveness in the domestic political round[9] was not matched by the realities of foreign aid policy, where the MFA often led in domestic policy-making.

The formal tasks of each ministry were defined in ministry organisational ordinances, although usually in vague and legalistic language. As can be seen from Figure 3.2, the MFA Economic Cooperation Bureau was the largest aid section in any of the ministries. It had seven divisions and dealt with most aspects of foreign aid policy: loans, grants, technical cooperation and multilateral aid. It also had a brief for general policy planning, although in reality that was possible only within the constraints of Japan's single-year budgeting. The power of the MFA in aid policy sprang from its broad authority, its ability to gain an overview of aid policy as formulated and implemented, and from its management of Japan's foreign relations, of which aid was a vital, and towards the developing countries often the most telling, component. It was the official 'window' for all Japanese aid business, a function which it jealously guarded.

Other ministries positioned their aid divisions closer to the primary policy bureaus. MITI's Economic Cooperation Department, part of the International Trade Policy Bureau, was intimately associated with the regional and bilateral trade policy areas of the bureau. The department's two divisions were concerned with economic and technical cooperation respectively, as they affected trade, which meant that MITI's aid work overlapped that of the MFA in two domains, technical

Figure 3.2: The Structure of Japan's Aid Administration, 1978: The Main Ministries

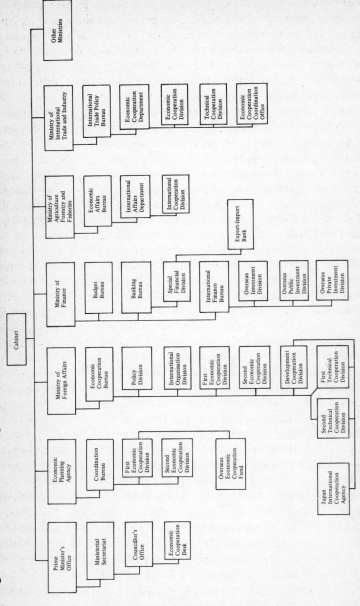

Source: Administrative Management Agency.

aid and loans. MITI remained influential in aid policy because of the continued relevance of the past, since aid and commercial policy had been closely interrelated for twenty years. MITI had become one of the largest and most powerful of the domestic ministries because of its direction of Japan's postwar industrial and trade growth; this strength spilled over into aid policy at several points, especially as loans often involved industrial or resources projects.

This system of 'equal partnership' as some officials described it, was a delicate arrangement of checks and balances, where formal jurisdiction was not always matched by real authority, where the MFA relied on its experience in international affairs to offset the weaknesses in its domestic power base, and where a Ministry of Trade was at the same time a Ministry of Industry, using both sets of (sometimes conflicting) powers in domestic policy-making. Likewise, the MOF's policy sword was double-edged, dealing both with aid in the budget context and with aid as a tool of international monetary policy. The three Overseas Investment Divisions in its International Finance Bureau had jurisdiction in government loans, multilateral aid and private investment. The Budget Bureau contained an Economic Cooperation and Foreign Affairs Desk which assessed all budget requests for aid, while the Banking and Financial Bureaus were also involved in supervising Eximbank and OECF affairs.[10]

Other domestically powerful ministries, such as the MAFF, had an increasing interest in aid policy (as the JICA study revealed), but the MAFF's aid function was restricted to technical aid and did not extend to capital assistance, except for JICA loans. In 1979 the MAFF was still excluded from formal participation in the four-ministry committee deciding on loans policy. Aid to agriculture had for a long time been a small part of the government programme, partly because of lack of support from within MAFF: until the early 1970s aid had been regarded as peripheral to the first objective of the ministry, promoting domestic agriculture.

The other main actor, the EPA, was a minor force in the bureaucracy. Although the architect of the national economic plans, which had guided economic policy since the late 1950s and had allied aid loosely with export and foreign investment goals, the EPA participated in the aid administration because of its responsibility for the OECF. Its Coordination Bureau included two Economic Cooperation Divisions. The main task of the divisions was on paper enormous, the 'general coordination of aid policy', but few took their contribution seriously. The EPA, it was said, had no 'voice', no philosophy of aid or concrete resources for policy-making. Any attempts at coordination, as will be explained, were weakened by the ambiguous relationship between foreign aid and national priorities and confusion between the competing components of the aid policy package.

The weakness of the EPA in policy discussion was partly attributable to controls exercised by other ministries, especially the MOF and MITI, through a long-standing system of transfer appointments of ministry officers to select EPA positions. While MITI and the MOF vied for influence by placing their officials in the highest posts, lower down — in the Coordination Bureau for example — the Bureau Director usually came from MITI, the Councillor (*sanjikan*) from the MOF and the Coordination Officer (*chōseikan*) from MITI, while the Division Directors were MITI (in the First Economic Cooperation Division), MOF or native EPA officers. Transferred officers, typically career officers from the larger ministries gaining outside experience as part of their career development, also occupied positions below the division director level.[11]

Aid Within Ministries

As aid policy was accorded fitful recognition by the national bureaucratic structure, so aid sections were regarded differently within each ministry. The standing of bureaus within any ministry changed periodically with the fortunes and importance of the duties which they administered, but the status of the Economic Cooperation Bureau in the MFA, for example, was never fixed. General perceptions of the importance of the developing countries in foreign policy were transmitted to aid, but in the MFA the European and American spheres were still the popular diplomatic domains. While it was acknowledged that aid experience was helpful to an officer, the prestigious diplomatic postings were usually not seen to be in developing countries, although this appeared to be slowly changing in the late 1970s with the further multilateralising of Japan's foreign policy interests.

The Economic Cooperation Bureau was still something of a backwater in the MFA in the late 1970s and, while the government's aid initiatives of 1977-8 added to the bureau's workload and responsibility, ministry customs took time to adapt. Established in 1962, it was considered a 'new' bureau, not one of the traditional branches of MFA structure, and it did not greatly impinge on power relations within the ministry. No Economic Cooperation Bureau Director had risen to the ministry's top post, the vice-ministership, possibly because of the country 'factions' identified by some writers as important in intra-ministry power struggles;[12] the generalist traditions of the MFA certainly did not encourage functional allegiances. In the late 1960s, when Japanese aid began to grow rapidly, some junior officers looked upon the Economic Cooperation Bureau as a useful base for rapid promotion.[13] The bureau was, however, a functional one, to many merely for implementation rather than for policy-making, and ministry custom perpetuated generalist thinking towards careers in the regional policy bureaus. There seemed to be no conscious practice of posting bureau officers to developing-country embassies, or vice versa.[14] There were, as will be shown later, few aid specialists in the MFA and for certain structural reasons there was no attempt to develop an aid 'service'.

The dichotomy between policy-making at the centre and work 'on the ground' was implicit in the reluctance to specialise. The distinction appeared often in Japanese administrative studies, in relation to the head office/branch division in business firms or the links between the MFA and its embassies.[15] In the aid context, the ministries and the implementing agencies represented the two poles, and Tendler noted the same phenomenon within a single aid agency, the United States Agency for International Development (AID). Nakane has implied that because of its socio-anthropological roots the dichotomy is a permanent feature of large organisations; whether or not that is so, the short-term implications for the management of aid policy were obvious: in general terms there was a preference for the Tokyo end of the policy process, a reluctance to be involved with the bricks and mortar of aid projects, and a view of service in the aid divisions as a phase in one's career, not a career in itself. The approaches to aid tasks of, for example, OECF and MFA officials, was remarkably divergent, the former typically aggressive in their arguing of the case for aid specialists.

The importance of aid sections within MITI and the MOF was expressed in terms of administrative capability, perhaps in keeping with the self-conception of those ministries as more practical and realistic. MITI's aid divisions were called 'special administrative units', possessing both strong intellectual and administrative resources. It was acknowledged that division directors in the Economic Cooperation Department were men with bright career prospects, and it was said that being in the department was conducive to an officer's promotion. Strength was needed in an extremely busy department which coordinated all aid-related business in the ministry. One of the department directors since

1966 later became Vice-Minister of MITI (Yamashita Hideaki in 1973). Some became senior bureau directors (Takahashi Toshio and Moriyama Shingo) while others moved to top positions in the Export-Import Bank (Yamaguchi Eiichi) and the Science and Technology Agency (Koyama Minoru). Former division directors included Harayama Yoshifumi, later a director of Dengen Kaihatsu, a semi-government engineering development firm, Eguchi Hiromichi (later a bureau chief in the Defence Agency), Hanaoka Shūsuke (Industrial Bureau Director in the Osaka Branch of the Ministry), Toyoshima Tadashi (Director of the General Affairs Division of the International Trade Policy Bureau and then a department director in the Resources and Energy Agency) and Kōno Kenichirō (Director of the General Affairs Division of the Industrial Location and Environmental Protection Bureau).

Differences between bureaus within a ministry were more noticeable in the MOF, where the International Finance Bureau had long had the reputation of something of an outsider. The international face of a closely knit and inward-looking ministry, it was a 'foreign cadre' which had always been considered somewhat special. Bureau consciousness was aggressive, despite regular movement of officials, and the bureau was staffed with capable officers. This was put down not only to the inherent difficulties of managing international finance, but to the need for staff well versed in foreign languages and able to manage Japan's economic diplomacy.[16]

The size and rate of growth of aid divisions also reflect their administrative standing, although size alone does not suffice as a measure of quality. Japanese national government operates on the *teiin* system (a fixed personnel or type of 'staff ceiling' system) as the means of determining staff levels. The annual budget designates numbers per ministry of executive officials, and total civil service staff numbers are set in special legislation. Ministry orders then specify bureau ceilings while the allocation of staff between divisions is decided by the Minister's Secretariat of each ministry after discussions with the General Affairs Divisions of the bureaus. Staff in divisions are shifted according to changes in divisional responsibilities and workloads.

Table 3.1 shows fixed numbers of employees (*teiin*) of the various bureaus for the period 1965-75 and actual numbers of employees in aid divisions within the ministries. The surprising fact that emerges from even these indicative figures is that in the ten-year period there was no real expansion in the aid staff of the main ministries, even though the funds allocated to aid — and the workload — expanded enormously. Actual numbers in MITI and the MOF showed no increase over the

Table 3.1: Staff of Aid Divisions in Main Ministries, 1965-75

	EPA			MOF				MITI							MFA						
	Col. 1	Col. 2	Col. 3	Col. 4	Col. 5	Col. 6	Col. 7	Col. 8	Col. 9	Col. 10	Col. 11	Col. 12	Col. 13	Col. 14	Col. 15	Col. 16	Col. 17	Col. 18	Col. 19	Col. 20	Col. 21
	(f)	(f)	(f)	(a)	(a)	(a)	(f)	(a)	(a)	(a)	(f)	(f)	(f)	(f)	(f)	(f)	(f)	(f)	(f)	(f)	(f)
1965	8		52	15	14		159	15	16	15	279	2	10	8	13	24	23			13	93
1966	8		52	11	10	13	161	15	15	14	262	2	10	8	15	23	23			12	93
1967	8		52	12	12	13	164	19	17	13	260	2	10	8	15	21	23			11	90
1968	8		52	13	13	13	164	18	17	13	260	2	10	8	15	21	23			11	90
1969	8		51	13	12	12	162	18	16	14	257	3	10	8	14	18	23			10	86
1970	7		50	15	12	12	161	17	17	13	257	3	10	8	14	18	23			10	86
1971	7		50	15	10	12	159	18	16	13	255	3	11	8	14	18	23			10	87
1972	7	6	53	16	12	12	159	18	16	13	253	3	11	8	14	22	18	9			85
1973	7	6	54	15	12	12	157	24		13	194	3	11	8	14	22	20	9			87
1974	9	6	60	15	12	14	155	26		12	195	3	13	8	15	22	20	10			91
1975	9	6	59	12	12	14	155	26		12	195	3	13	8	15	19	18	10	8		94

(a) refers to actual numbers of staff (jitsuin).
(f) refers to fixed numbers of staff (teiin), also called 'establishment'.

EPA: division numbers were taken from agency sources and were as of 1 April (fixed by agency policy). Bureau numbers were taken from Jinjiin kanrikyoku shokkaika, Gyōsei kikan soshikizu (National Personnel Authority, Bureau of Administrative Services, Position-Classification Division, Organisational charts of administrative organs) various years, and were as of 1 July.

MOF: division numbers were derived from ministry sources and were as of 1 April. Bureau numbers were taken from Gyōsei kikan soshikizu, various years, and were as of 1 July.

MITI: division numbers were from ministry sources, and were as of 1 April while bureau numbers were taken from Gyōsei kikan soshikizu, various years, and were as of 1 July. 'Bureau' refers to Export Promotion Bureau 1965-72 and International Trade Policy Bureau, 1973-5.

MFA: all numbers derived from Gaimushō keizai kyōryokukyoku, Kikō oyobi teiin no hensen (Ministry of Foreign Affairs, Economic Cooperation Bureau, Changes in structure and personnel), April 1976, and were as of 1 April. Numbers of divisional staff were fixed by ministry policy.

Key

Col. 1 – Economic Cooperation 1 Division
Col. 2 – Economic Cooperation 2 Division
Col. 3 – Coordination Bureau Total
Col. 4 – Overseas Investment Division
Col. 5 – Overseas Public Investment Division
Col. 6 – Overseas Private Investment Division
Col. 7 – International Finance Bureau Total
Col. 8 – Economic Cooperation/Policy Division
Col. 9 – Capital Cooperation Division
Col. 10 – Technical Cooperation Division
Col. 11 – Bureau Total
Col. 12 – Bureau Director/Counsellors
Col. 13 – Policy Division
Col. 14 – International Organisations Division
Col. 15 – Economic Cooperation 1 Division
Col. 16 – Economic Cooperation 2 Division
Col. 17 – Technical Cooperation 1 Division
Col. 18 – Technical Cooperation 2 Division
Col. 20 – Reparations Division
Col. 21 – Bureau Total

decade (in MITI they in fact fell) and fixed bureau numbers dropped in both cases. In contrast, the EPA levels increased slightly and the creation of a new division in 1972 did not, as was common in such reorganisations, draw staff from the First Division.

Figures for MFA staff numbers, while for *teiin* only, do reveal something of the formal division of workloads within the Economic Cooperation Bureau. Employees in divisions whose functions did not greatly change (Policy, International, First Economic Cooperation) showed a slight increase over the period, as did the number of technical cooperation staff, which benefited from a new division in 1972. Grant aid and reparations staff decreased as reparations were wound down and people transferred to other tasks. Bureau fixed numbers stayed almost the same, while regional bureaus and the Minister's Secretariat substantially increased their personnel, especially after the reorganisation of 1969 which transferred to the regional bureaus some of the responsibility for bilateral economic relations previously residing with the Economic Affairs and Economic Cooperation Bureaus. In particular, in the other main functional bureau, Economic Affairs, *teiin* dropped from 178 in 1968 to 88 in 1969, while the Economic Cooperation Bureau lost only four of its 90 positions. Regional bureaus had their numbers increased by between 13 and 46 per cent. By 1978, Economic Cooperation Bureau *teiin* had increased by two over 1975, although this masked a further fall in grant aid staff of three (due to the transferring of some work to JICA), with rises in four divisions.

Apart from the generally low political interest in aid in Japan, one factor which helped account for the static numbers of aid staff was the policy of the Administrative Management Agency and National Personnel Authority to restrain growth in the size and manpower of the bureaucracy as a whole. This was effected by across-the-board reductions of staff, and even of divisions. In three programmes carried out between 1968 and 1975, 98,000 personnel (*teiin*) were cut from existing divisions, while 86,000 posts were created for new administrative areas, making a net reduction of 12,000.[17] This occurred together with regular rationalising of staff within the bureaucracy as functions changed, a process which was encouraged by the annual review of staff levels in the budget context. This environment for continual assessment of personnel levels was not simply a result of budgetary pressures; since the work of the Temporary Commission on Administration in the mid-1960s, advisory bodies like the Administrative Management Committee had pressed for strong measures in controlling bureaucratic growth.

Table 3.2: Foreign Ministry Economic Cooperation Bureau: Actual Staff Numbers as of 1 April 1976

Division	Fixed Framework			Outside Fixed Framework				1976 Budget Staff Levels	Actual in Fixed Framework
	MFA	Attached to Bureau	Other Ministries	Other Ministries	Private Companies	Temporary Staff	Total		
Bureau Director	4					1	5	3	4
Policy	15		3	1	1	2	22	13	18
International	8		1		2	1	12	9	9
Economic Cooperation 1	14		1	2	1		18	15	15
Economic Cooperation 2	13		3		1	1	18	19	16
Technical Cooperation 1	16			1	4	1	22	18	16
Technical Cooperation 2	13				1	2	16	10	13
Development Cooperation	6		3	1	3	2	15	8	9
Reparations Office		3	2			1	6	5	5
Total	89	3	13	5	13	11	134	95	105

Fixed Framework (teiin taishō): Foreign Ministry pays the salaries of staff from other ministries in this category.
Outside Fixed Framework: staff seconded from other ministries, companies etc., for varying periods. Temporary staff are female secretaries.
Source: Gaimushō keizai kyōryokukyoku, *Genin shirabe,* (Ministry of Foreign Affairs, Economic Cooperation Bureau, *Staff survey*), April 1976.

Fixed staff figures indicate the number of positions allowed within the budget framework but, as Table 3.2 reveals, for example, there was considerable flexibility in staffing, for personnel in the MFA's Economic Cooperation Bureau in April 1976 totalled 134, 39 more than the number allocated in the budget. From the table we can see that MFA's Economic Cooperation Bureau was not an isolated or 'ethnocentric' bureau. Its fixed staff included 13 transferred from other ministries (Finance, Construction, Agriculture and Forestry, Home Affairs, Transport, Posts and Telecommunications), while officials were also seconded from ministries, banks and semi-government organisations. A number of these were technical specialists. The large discrepancy between fixed and actual staff numbers was regarded as a normal state of affairs in the bureau, although the total was probably changing constantly. The dearth of aid specialists in the MFA may have been one good reason for the considerable borrowing of staff.

Other ministries did not appear to have staff much in excess of the figures quoted in Table 3.1 but, within the MFA's Economic Cooperation Bureau, all divisions had outside personnel. Only the grant aid division, however (the Second Economic Cooperation Division), had a total lower than its fixed allocation (18 as against 19), a fact which did not altogether tally with forbidding descriptions of the division's duties (see Chapter 5). The Development Cooperation Division was the best equipped, having 15 in comparison to a fixed allocation of eight. This reflected the emphasis of policy on JICA and on its feasibility survey work, the supervision of which the division was established expressly to carry out. The Policy Division, the most important in the bureau in respect of budgeting and the control of policy directions, had 22 staff members, but was matched by the 22 in the First Technical Cooperation Division which managed overseas technical assistance policy. Nowhere, however, was there a unit sufficiently large to attempt bureau coordination, much less overall coordination of foreign aid. This situation had not changed in 1978. Neither were there sufficient personnel in the Policy Division to permit the development of detailed country programmes, a task to which the United State Agency for International Development, for example, devoted vast resources.[18]

Other ministries did not possess the administrative muscle necessary for broadly based coordination. The EPA, which had a brief for it, only had an aid staff of 16 in 1976. The Councillors' Office of the Prime Minister's Office, which was suggested as an appropriate coordinator of aid policy in its role as Secretariat to the Advisory Council, was even smaller, with only four men on the Economic Cooperation Desk. The

International Cooperation Division of the MAF's International Department was very large, with 49 persons, but many of these were attached technical and other officers working on non-aid matters. According to Table 3.1 (see p. 93 above), aid-related staff in the three main ministries and the EPA totalled in 1975 only 185, whereas in 1965 they numbered 176. Even taking into account extra personnel recruited from outside, staff resources over the decade after 1965 nowhere matched the increasing workloads represented by aid performance figures, table staff numbers, reinforced by civil service strictures against specialisation, widened this gap between ministry workloads and their capacity to meet new demands. As was pointed out in Part I, the aid administration changed little in outward structure or division of responsibilities after the mid-1960s, while the aid programme grew and diversified rapidly. It is now clear that internal structures did not fully respond to these changes, for various reasons, and that the process of adjusting is still going on. One result (as Part III explains) was that much of the task of recipient contact and project management was diverted to the implementing agencies (OECF, OTCA and later JICA) and other non-government actors, a trend which had profound implications for the pattern of aid-giving.

Aid and Bureaucratic Behaviour

Aid did not exert a powerful bureaucratic influence in Japan; it was peripheral to the main purposes of the ministries which were an active part of the aid system, and tended to fall neatly between the two stools of political and trade relations with individual countries. It drew only sporadic support as a set of policies *per se*, while particular aspects of the work of aid divisions were more likely to appeal to external interests. This dichotomy between the whole programme and its parts severely restricted the capacity of divisions to assess their work in the light of more general trends in the aid programme. At the same time, the daily operation of divisions placed more immediate limits on the aid programme, since aid was managed by small, disparate groups of officials, between which the barriers were predominantly vertical.

Traditionally, as representatives of the Emperor and of state power, officials were set above the mass of Japanese society, secure in prestige and social position and surrounded by an aura of administrative legalism. In postwar years, however, the Japanese bureaucracy became extraordinarily open in its methods, far more so in fact than many

Western systems. The was close and constant communication between administrators and their clients (taken too readily by some commentators as sufficient evidence of Japanese-style collaborative policy-making) and a correspondingly heavy traffic of officials and the public in and out of ministries every day. This reflected a social dependence on bureaucracy, a need for official approval and licensing of much daily business,[19] as well as certain cultural values which emphasised face-to-face communication as an expression of formality, ranking and procedure in social contact between primary groups. These social norms contrasted with patterns in countries like France, where Crozier detected a preference for authority relationships and for systems in which impersonal interoffice memoranda replaced direct communication. For the highly decentralised Japanese foreign aid bureaucracy, as in other areas of government, the emphasis on face-to-face communication and personal ties assisted coordination; it helped people know what others were doing and thinking.[20]

The lowest unit of formal organisation and the focus of routine policy-making, the division (*ka*) within a ministry, exhibited these characteristics most clearly. Authority within the division was vested in the director (*kachō*), his deputy directors (*kachō hosa*) and in the various desk heads (*kakarichō*), but the division's small size (usually 10-20) meant that the director himself could maintain constant contact with his officers. A division's basic office layout rarely varied and structures of authority were reflected in the arrangement of desks, the director usually sitting by the window across the corner or in the centre facing the desks of junior officers and flanked by his deputies. The accent on the functional unity of the division gave no concession to privacy, and group consciousness (even camaraderie) predominated, consistent with patterns of Japanese social behaviour. Because the division was a close unit, effective management was necessary for the completion of tasks, and the personality of the director was decisive in translating authority into collective action.[21] Communication between ranks within the division was easy, frank and constructive, because authority was not expressed, as Crozier discovered was the case in France, in rigidly isolated strata and the linking of separate ranks by regulations and orders, but rather in a creative energy which enveloped the group. The individual was consciously part of a team which allowed him to participate informally in decisions often outside the scope of his immediate responsibilities, but which left him very little room for independent action or initiative.

Despite stable authority structures within the division, however, harmony and consensus were not all-pervasive. The division was the focal point for several underlying tensions which cut across integrative patterns of authority. The career system was one such factor which, while providing goals and expectations for the individual, threatened the order of the division. In relation to foreign aid, for example, career patterns had a distinct impact on policy-making, in separating the 'elite' stream of officials from any concept of specialisation.

Entry to the Higher National Civil Service and advancement within it were determined ordinarily by examination. The system of upper, middle and lower categories of examination operated as a streaming mechanism, channelling personnel into distinct career structures, where there were (within certain margins) fixed upper limits to promotion and uniform rates of upward mobility. The numbers who passed the Higher National Civil Service Examination and entered the ranks of the upper stream (termed the 'career' group) were very few. For example, in 1973 they numbered only 639 out of 10,826 who passed examination in all three categories.[22]

Both career and non-career streams comprised generalist administrators and technical officers. Only those who were successful in the Higher National Civil Service or the Higher Foreign Service Examinations were eligible to become the highest officials in the national bureaucracy and so, in the ministries (at least those which dealt with aid), senior officers (division director and above) were usually generalist career officers who had moved upward at a fixed and steady rate,[23] with little open competition for advancement between them until the number of candidates began to outstrip available posts at the division director level. Career officers were promoted much faster than were non-career officers and changed positions every two years or so; they moved from bureau to bureau and often to attached agencies or even to other ministries (as noted in the case of the EPA or transferees to the MFA's Economic Cooperation Bureau), gaining wide and varied training in administrative affairs. Non-career staff, however, advanced slowly and spent long years in the same job in the knowledge that they would never achieve senior positions of power in the ministry, but at the very most might expect to be given a directorship of a minor division after 20 or 30 years' service.

These distinctions created tensions in the workplace and frustration among the ranks of non-career officers at being isolated from higher levels of policy, particularly when they acquired specialised knowledge of the policy area in which they worked. There was, in fact, a severe imbalance between career structures and the need for expertise, of which foreign aid provided a good example. Career officers received a very general training and often relied on non-career personnel for the knowledge of rules, precedents and internal divisional guidelines which was essential to much of the work of the bureaucracy.[24] Because career officers moved regularly, divisions needed a stable reservoir of experience. This was provided mainly by non-careerists. The small numbers of career officers in ministries exacerbated problems of generalism, because effective deployment of career officers became very difficult and resulted in their being spread thinly across a ministry. In the case of the MFA, Fukui has shown how this resulted in 'single-issueism' and policy-making by improvisation.[25]

The continuity of foreign aid policy was seriously affected by the exclusion of specialists from what were careerist posts. As Cunningham showed, specialists had to be active in the central management branch of aid agencies if their skills were to be fully exploited.[26] In Japan, however, technical experts were grafted uneasily onto the system and non-career officers, the most experienced lower-level administrators in the ministries, were separated from responsibility for policy decisions by inflexible and traditional barriers. The Japanese system did not cultivate aid specialists, and the single research officer appointed in 1975 to the MFA Economic Cooperation Bureau's Policy Division was the first career officer in what was normally a non-career officer's final appointment. MITI appointed a 'specialist officer' in 1976 but he was non-career. The personnel system prevented specialist aid officers being promoted to a high level in the ministries, a problem which other donors (such as Australia) also faced.[27] On the other hand, the agencies — JICA and the OECF — had staffs which typically spent their careers within the same organisation and officers with long experiences in aid business, in Japan and in the field, could rise to senior positions.

In practice, the extent of face-to-face contact in much of a division's daily work, while reflecting the need for separate organisations to communicate horizontally at a personal level, resulted in an apparent informality and compensated in part for rigidity imposed by organisation and formal regulation. In the case of the MOF and MITI, for example, jurisdiction was defined in great detail and was, for all ministries and their bureaus, a source of perennial argument. Some would say that it was the overriding concern of officials, particularly where, as one officer of the MFA's Economic Cooperation Bureau described it, 'the broad scope of our work is clear, but the dividing lines in difficult issues are exceedingly complex'.[28] Informal and consensus-oriented processes within the division also acted as a balance to the permanent status divisions between officers.

More importantly for decision-making, the emphasis on the division as a group meant that the scope for initiative was limited to the extent which the group would allow. This was quite different to the innovative behaviour of AID lower-level and field officers which Tendler discussed.[29] In the absence of a higher directive, the division moved cautiously and reacted slowly to non-routine problems especially when these demanded coordinating with other divisions. The need specifically to extract officers from the divisional environment to make up task forces to tackle crises or irregular situations was well known: the JICA case was typical.[30] The ease of vertical relationships within the division contrasted with the problems of horizontal relationships between divisions, which were manifested in the face-to-face encounter as a means of bridging boundaries. The unity of the division was derived from its group orientation and from its being the lowest organisational unit with legally defined existence and responsibility. The workplace environment in fact encouraged dispute over policy at the lowest ranks. Jurisdiction became controversial from the bottom up, as argument over substance became argument about territory, and as the psychological commitment to the home division was expressed in policy terms. When energies were devoted in this way to maintaining current group goals, innovative behaviour was typically the result of external stimuli, or was due to inputs from sources outside the bureaucracy.

Advisory Bodies

One possible source of such initiatives or of guidance for interdivisional tasks was the public advisory body. Common in many countries, including Japan, as a vehicle for outside advice to governments, it displayed functions which, according to Harari, were both instrumental and systemic.[31] In policy guidance, legitimising government action or inaction or co-opting of interest groups into the policy-making process, advisory bodies were always characteristic of Japanese government, which boasted some 244 in 1978. Foreign aid was one area where these committees were used. A number of aid donor countries established permanent advisory boards to advise their governments on aid and related subjects. In Australia, for instance, the Development Assistance Advisory Board was created in 1974 to advise the Minister for Foreign Affairs, but ceased to exist in June 1977 when the Australian Development Assistance Agency was reconstituted as a bureau under the control of the Department of Foreign Affairs. Such bodies also existed in Austria, Belgium, Denmark, Finland and the Netherlands and all had both government and outside membership except those in Denmark and the Netherlands, which were composed entirely of people from outside government.[32]

The Japanese government's advisory body on foreign aid was the Advisory Council on Overseas Economic Cooperation (*Taigai keizai kyōryoku shingikai*), first established in 1960 and restructured in 1969 to achieve a greater non-official membership and, it was hoped at the time, a livelier contribution to the national aid debate. Once a group of ministers which never convened, it became a council with a top-ranking businessman (Nagano Shigeo, Chairman of the Japan Chamber of Commerce) as chairman and a leading academic and former bureaucrat (Ōkita Saburō, then President of the Japan Economic Research Center) as deputy chairman. Its resurrection in 1969 followed a decision in September 1969 by an *ad hoc* Ministerial Committee on Economic Cooperation formed in the previous May. The reasons for the Council's reorganisation were twofold: to achieve guidelines for Japan's aid policies in the new international situation of the 1970s; and to answer widespread criticism of the inadequacies of Japan's aid policies and administration. The *ad hoc* Ministerial Committee was intended to supplement the council but in fact rarely met and, after recommendations from the council in May 1975 and pressure from within the LDP, it was upgraded to an official Ministerial Committee on Overseas Economic Cooperation in July 1975.

The council was an advisory body to the Prime Minister and was, in most respects, typical of other advisory bodies. It exhibited the same weaknesses and strengths which the literature has amply described.[33] The council was a small group, having only 20 members, but it represented a cross-section of academic, business and semi-government groups, as was typical of advisory commissions in Japan. The number of present and former officials was always large and tenure was for the usual two years. Meetings were held regularly, about once a month, and representatives from the ministries and, on occasion, officials from implementing agencies, attended the meetings. Subcommittees were established when necessary and drafting of reports was undertaken by a drafting committee. Reports were presented to the responsible minister (in this case the Prime Minister) in the customary way.

While the structure and composition of the council were standard, the relationship between it and the policy area upon which it was meant to advise, was unusual. It made recommendations to the Prime

Minister because it was always attached to the Prime Minister's Office and because in 1960 there was no one minister predominant in aid policy whom it could more suitably advise. This was still the case in the late 1970s, so that while legally the council was in a neutral bureaucratic position, it was also separated from direct contact with any one ministry involved with aid. This weakened the council's influence over the fate of its reports, and meant that the council had no 'promoter'. Thus, in the absence of a 'promoter', the council did not convene at all for the first six months of 1975 or 1976, not even to discuss policy for UNCTAD IV, which began in May 1976. The Prime Minister was concerned with aid policy on a high and abstract plane and could not be counted on to push for the adoption of council recommendations unless he was personally committed. Both he *and* the council were removed from the daily round of aid policy, but even though the council's organisational position isolated it from the centre of policy development in ministries, its reports still had a stimulatory effect and could be a source of new ideas for the government. At the very least it could act as a sounding board for initiatives from the bureaucracy or government at large.

Most advisory councils were, to some extent, controlled by the officials of the ministry to which they were attached. This control could emasculate a potentially active council or, as was the case with MITI's Industrial Structure Council (which advised and reported on Japan's industrial structure plans) or with the EPA's Economic Council (which drafted the nation's economic plans), it could add legitimacy to policies which originated in the ministry itself. The reports of the latter two councils, for instance, were immediately adopted as policy and the councils were correspondingly accorded the status and outward signs of power consistent with their role as legitimising agents.

The Advisory Council on Overseas Economic Cooperation performed a similar function, but it was weak and of low status, due partly to long years of inactivity. Its rebirth in 1969 failed to pull it out of administrative limbo and no pattern of policy leadership emerged. The Council Secretariat was the small Economic Cooperation Desk in the Councillors' Office of the Prime Minister's Office, and none of its staff were specialists, in contrast to the unit serving MITI's Industrial Structure Council. As a result, there was no possibility that the desk could attempt to coordinate aid policy, unless it was enlarged and its powers expanded. The council, in its own report of August 1975, recommended that strengthening of the Council Secretariat be considered, although officials prevented a positive wording in the draft of the report. Ministries did not want the secretariat's staff or its coordinating ability upgraded and there were even half-hearted suggestions that the secretariat should be moved from the Prime Minister's Office to either the MFA or the EPA.[34]

Council deliberations were dominated by officials who were present at both general and subcommittee meetings to give briefings and to observe. They often outnumbered councillors and sometimes assembled from 15 to 16 government ministries and agencies. The format of the meeting, which was fairly formal and of only a few hours' duration, constrained councillors in freely discussing the items on notice, so they relied on circulated papers to direct debate and often reserved their comments for written submissions made outside the council meeting. The Economic Cooperation Desk in the Prime Minister's Office was responsible for distributing papers and most of the agenda was supported by ministry-prepared documents. Many of these position papers were so contradictory that some councillors were moved to suggest in subcommittee that ministries do more to coordinate their drafting. Officials did not form any bloc in their approaches to the council. Discussion at general meetings was often lively and members were free to raise topics other than those on the prepared agenda. Officials participated in debate, frequently without being asked their opinion by the chairman. They were, therefore, more than observers and did not hesitate to disagree with councillors, who themselves displayed an extremely wide range of opinions about foreign aid and about the system of aid management. Regularly the discussion became a mere dialogue between officials and councillors rather than an independent, constructive debate between council members as such.

The efficacy of the Advisory Council can be measured best by the content, and impact, of its recommendations. These were prepared by a drafting committee and were based on subcommittee findings and on other documents arising from council meetings. Drafting was concentrated in the Economic Cooperation Desk, for this office collected and edited opinions on drafts from councillors and from ministries. The influence of bureaucrats on the final composition of drafts is not easy to gauge, but it depended very much on the chairman of the drafting committee. Influence was obviously a two-way process, although some critics of the council accused it of a 'scissors and paste' exercise in combining different ministry papers to produce a report. Some members complained that reports were foregone conclusions anyway, because of the strength of ministry influences.

Reports testified to the fact that the council was not so much directed by ministries as overwhelmed by the intensity of their differences. Since the capacity of the council to formulate its own independent papers was limited (no councillors had time, for a start), it necessarily relied on ministry material for the bulk of any draft. Ministries vied to get their proposals to the drafting committee quickly and forcefully. The form of the draft and the final report depended much on the ability of the drafting committee's chairman to counter the generalities imposed on him by the conflict of bureaucratic interests. The council had to tolerate generation of heat rather than light in this infighting and its overriding problem therefore was how to ensure that its recommendations were properly implemented. In this, however, it was dependent on the whim of officials, something which the isolation of the council from the ministries only intensified.

One appraisal of the implementation of the September 1971 report on technical cooperation (which suggested reforms in practically all areas of Japan's technical aid programme) revealed that, in the complex and highly specific field of policy, implementation was overseen by twelve ministries and agencies.[35] This facilitated improvement of a few small, isolated programmes (such as Ministry of Education scholarships to foreign students) and there was some effort in respect of medical and agricultural cooperation, the Japan Overseas Cooperation Volunteers and project survey work, although there were other factors in these last two which pushed programmes along independently of council advice.[36] These successes constituted, however, a small proportion of total technical assistance and the important policy categories, where cross-ministry coordination was essential – such as the use of technical experts, training in Japan of LDC personnel, setting of policy guidelines and acceleration of the disbursement of technical assistance – met with little improvement.

In formal terms, the council sat at the apex of the aid administrative structure but in that position lay the root of its impotence. Being alienated from the central aid machinery opened the council to attack from all corners of the aid bureaucracy and left members with no sense of their own status. There was no concept of tangible interests associated with the council and it was difficult to appreciate where aid stood in national policy priorities. Likewise, the council did not have a firm yardstick for assessing submissions or the details of aid policy presented to it. While charged with the task of advising, the council itself lacked a common standard for its advice. The initiative in policy recommendations rested with the officials, and the council could not force its ideas upon anyone. It was, ultimately, only an advisory body and the ministries translated those ideas into concrete policy as, and if, they wished.

In contrast, the Ministerial Committee on Overseas Economic Cooperation was one of several groups of ministers set up by Cabinet decision to discuss issues with which Cabinet as a whole could not deal. Its creation in 1975 was prompted, according to news reports, by the need to achieve greater unity and 'hard' ministerial level coordination in aid policy and to rationalise private and official economic cooperation programmes. The Cabinet decision itself cited the need to draw up 'basic policy'.[37] At the same time, pressure from within the LDP was instrumental in bringing the question of a committee to the attention of the Cabinet. One report suggested that Deputy Prime Minister Fukuda first raised it in April 1975 in connection with attempts to increase aid to South Korea, negotiations about which had been suspended since the Korean Opposition leader, Kim Dae Jung, had been abducted from Tokyo in 1973.

If Fukuda did raise anything, however, it would have been proposals by an LDP member of the House of Representatives, Minato Tetsurō, made in a paper in October 1974 where he suggested reconstituting the *ad hoc* Ministerial Committee as a full ministerial committee, to be a 'control system' for aid policy. This idea originated in talks Minato had had with the Prime Minister, Tanaka Kakuei, in December 1973 on the establishment of the Japan International Cooperation Agency. The Opposition parties, through the Diet, were also effective in focusing attention on aid in a way which reinforced Minato's good timing, when the JSP put up a bill aiming to regulate the types of recipients of Japan's aid.

The committee used as staff the Cabinet Councillors' Office, formally a separate body to the Councillors' Office of the Prime Minister's Office but in fact housed in the same room and administered by the same officers. Apart from the ministers, bureau directors and their assistants, the Chief Cabinet Secretary and the Chairman of the LDP's Policy Affairs Research Council and the Special Committee on Overseas Economic Cooperation also attended the committee meetings. No regulations regarding the agenda of the committee were laid down, but initiation of agenda items rested ultimately with ministry officials. The committee met five times beween September 1975 and July 1976 but not at all between December 1975 and June 1976. It did not confer on Japan's UNCTAD policy in April 1976, but instead a smaller group of 'main ministers' from the ministerial committee was convened, perhaps to give themselves more freedom in discussion. While it could assist in the coordination of all aid policies, the committee's attention was occupied by only a few topics, especially by Japanese participation in large-scale development projects overseas. A decision by Prime Minister Fukuda in January 1977 to abolish all ministerial committees put an end to its short life.

The committee was never considered able to live up to its espoused aims. As a Cabinet subcommittee it was not a forum where the details of policy were discussed and ministers had to rely on briefings from their officials. It existed for the raising of issues, not for making decisions, which was the prerogative of full Cabinet. Problems would go up to the committee from the ministries for discussion and resolution, but this meant that only subjects already enjoying substantial agreement at the ministry level became agenda items. They were 'procedural', originating in the ministries, and ministers dutifully reflected the views of their officials, as briefed.

This inertia was inevitable, for in a fixed administrative system such as that for foreign aid, a Cabinet-based committee was powerless to coordinate, or to provide new guidelines for, policy unless it had a full independent staff, a situation which Cunningham found for various European donors also.[38] For this reason, the creation of the committee was referred to as a 'token gesture'. Nevertheless, it did help to stimulate ideas from within the bureaucracy for a short period and led to one initiative, an insurance scheme for Japanese contractors posting bonds on the construction of large-scale projects. The committee was effective, but not in terms of its original brief.

The council and the former committee represented the only machinery for policy advice and high-level policy coordination in the Japanese aid system, but their advisory and coordinating capacities were weak. Since they sat at the top of a diffuse and divided administration, their influence was negligible by their being isolated from ministry power structures (in the case of the council) and their inability to focus on general trends in aid policy (in the case of the committee). The only contribution of the committee, the bond insurance plan, stemmed from hard political and commercial judgements about the success of Japanese companies in international tendering. Ministries directed policy-making unless pressure from outside and above was sufficient to ensure that advisory body recommendations were put into practice.

The Diet, Cabinet and the Political Parties

The aid bureaucracy was inadequate as a policy coordinator, the efforts of the Advisory Council to set new directions proved fruitless and the Ministerial Committee did not exist for long enough to reconcile the diversity of official policy emphasis. Could the National Diet, as the legislative arm of government, provide a policy-making forum? After all, the Diet had formal control over major administrative changes, and over the budget, the source of the aid programme. In some areas — education, labour, health and foreign policy — it was often more than a mere legitimising agent for the executive.[39]

There are three main ways in which a legislature can control an aid programme: the authorisation of funds (both of overall volume and the timing of expenditure); the general 'watchdog' role carried out as part of legislative scrutiny of the executive; and the passing of 'charter' legislation for the aid programme as a whole.[40] In Japan, only the first of these powers was exercised over aid policy. The Diet did not possess a formal aid committee and showed little inclination to scrutinise aid policy or to lay down controlling legislation. Aid came within the purview of at least nine committees in each House,[41] but they discussed aid as part of other business, debate being restricted by the division of committee responsibilities along ministry lines.

Aid budgets were approved by the Diet in the course of the annual budgetary process and while there would be some debate on aid in the Budget Committee, there was no tradition of dissecting figures which were, in any case, buried in the mass of budget documents tabled. Extraction and analysis of figures were simply too onerous for members to spend time over. As Chapter 5 will explain, Diet authorisation was necessary for grants (technical assistance, capital grants, multilateral assistance) and for some allocations to the loan agencies. The Diet also had to authorise the carry-over of undisbursed funds but again these figures were included among the detail of the budget papers. There was no Diet control of loan commitments, as in some other DAC countries, although the passing of the budget represented approval of certain grant commitments. Compared to the explicit authorising powers of the United States Congress, the Japanese Diet's role was ill defined.

Aid was peripheral to the business of the Diet, because in Japan aid was not a vital political issue. The public expressed little awareness of aid and only one or two members in the whole Diet possessed any professional knowledge of the subject. Aid comprised less than 1 per cent of the Budget's General Account and members gained no votes and little kudos for championing a cause which did not directly affect the vast majority of their electors. Ministries, despite a latent feeling that some sort of legislative control might improve their international image as aid donors, did not wish to see any aid committee established in the Diet since it would only increase pressure on their own overtaxed staffs and, by implication, let the Diet watchdog off its leash. On the other hand, it would be wrong to give the impression that members of the Diet did not care about aid. There were members in all parties anxious to see some public accountability in foreign aid. They reluctantly accepted that ministries went largely unchallenged in aid policy, but as the size of the aid budget rose — notably as a result of ministerial promises to step up Japan's aid effort — political interest did as well. There have been regular calls over the years for Diet approval of an aid budget package, or Diet sanction of large bilateral agreements.

Aid and aid-related topics arose frequently in the Diet, which was one reason for the increased demands to establish some form of Diet aid committee, as a subcommittee of the Foreign Affairs Committee or a wider group on North-South issues. The Foreign, Budget, Audit and Trade and Industry Committees usually dealt with aid through legislation before the House which involved aid indirectly (such as revisions of the Export-Import Bank Law in the 65th Diet or the Export Insurance Law in the 63rd), through scrutiny of past expenditure on aid in the Audit Committee or by examination of certain bilateral relationships between Japan and developing countries. Debate was not well informed or incisive, and certainly did not approach the standard of debate on broader foreign policy questions which occurred more regularly. Questions or comments on the administration of aid arose periodically, as in the Audit Committee of the 51st Diet, in the Social Affairs-Labour Committee of the 63rd Diet on the strike of OTCA employees or in the debate on the JICA bill. Government backbench and Opposition members themselves bemoaned their own unpreparedness, but aid did not advance their political reputations.

What of Cabinet itself? Cabinet had powers which extended directly over aid. Not only did it ratify the government's budget proposals and approval formal matters going before the Diet, but it also approved aid agreements between Japan and the developing countries, which did not go to the Diet. Potentially, Cabinet possessed the executive authority to control the aid programme, but the problem was again that of the political significance of aid and its importance in the nation's political timetable. As with the former Ministerial Committee, Cabinet was a place for decisions, so the initiative lay with ministry officials in taking matters to Cabinet. The important decisions (on JICA, for example, or the size of the annual aid budget), were made independently of Cabinet. As Robert Ward has written,

> In sum, primary political power in contemporary Japan rests with the Cabinet, and in particular the prime minister; they probably perform more decision making of major importance than any other formal unit of government. They are, however, subject to constant interaction with and substantial influence from several official and unofficial groups. There is no simple answer to this question of political primacy.[42]

The political parties appreciated that aid was an element of foreign economic policy, but the consciousness of aid among them varied considerably. The LDP had the oldest aid committee, the Special Committee on Overseas Economic Cooperation established in 1959. It was one of the PARC's special investigation committees, membership of which was voluntary, and although the lowest level of party policy-making, it acted as the focus of official LDP aid work. The membership of the committee stood at about 70 and the chairman in 1978 was Noda Uichi, a long-standing member of the Diet and a former minister influential in Japan-Korea affairs. Other high-ranking LDP Diet members were members of the committee but only about ten were constantly active. Indeed, with the death in 1977 of Minato Tetsurō the committee lost perhaps its most experienced and enterprising member. The committee was said to meet once or twice a week on a range of problems, and its staff consisted of one elderly LDP official. The committee made few formal reports although papers and recommendations prepared by individual members or groups within the committee were issued from time to time with its sanction. Thus members included a group pressing for economic cooperation with Brazil and others working for greater emphasis on agricultural aid, in 1973 over the new agency and later. The committee did not produce by itself any party policy. It was a minor section of the LDP organisation and had no real voice in party affairs. What weight it could exert in policy resulted from individual members pursuing their own objectives. It was constantly referred to as a 'support group' for aid, a back-up organisation or 'mood-builder' for ministry policy.

The committee's support did not extend to policy in a broad sense, although it was a clearing house for ideas and arguments advanced in its frequent meetings with ministry aid officials. Its energies, or rather those of its members, were channelled more into particular aspects of aid, especially at budget time or in relation to loans projects or bilateral relationships. This feature will be examined in later chapters, for it drew strong criticism from Opposition politicians and from the public. Individual LDP members repeatedly became involved in and identified with particular aid relationships and the Special Committee provided an opportunity for these connections to be initiated. The interest of some members of the LDP committee in large and expensive projects overseas to the exclusion of what were, to officials, more pressing issues, prompted many remarks about the pursuit of personal before national interests. By way of contrast, as will be shown in Chapter 5, the influence of the committee in the budgetary process could be most helpful to ministries seeking higher allocations. As in any policy process, the interaction between officials and politicians always flowed in two directions.

Opposition parties also maintained aid or economic cooperation committees, although none was as active or as informed as that of the LDP. The Democratic Socialist Party (DSP) kept a watching brief over aid through its International Division and did so with some verve, in contrast to the Japan Socialist Party (JSP), which presented a vague and disordered view of aid in its publications. None of the Opposition parties was fully aware of the details of aid organisation although some wits would maintain that neither was the government. All parties included statements on economic cooperation in their policy handbooks, but only the LDP and the DSP had a reasonably detailed treatment of the subject.[43]

The Opposition used its limited resources to good effect in the Diet on occasion. A bill presented in January 1975 by the JSP Diet member, Den Hideo, sharply drew the government's attention to aid problems. The bill provided for Diet approval of proposed government aid programmes and for the cessation of aid to non-democratic governments, a provision aimed specifically at South Korea. While it was never debated, the bill did coincide with Minato's proposals for a ministerial aid committee and with attempts by Fukuda to recommence aid talks with the South Korean Government.

Opposition members were diligent in attacking the government over foreign aid (as it provided a useful peg on which to hang rumours of financial scandal), albeit within the limits of parliamentary timetable and procedure. Reparations was one target of these thrusts, and one of the most effective was that concerning the alleged corruption in Indonesian reparations, first raised by Yanagida Hidekazu of the JSP in 1959. Questions were asked about the propriety of only a few Japanese companies (notably Kinoshita Sanshō and Nippon Kōei) gaining extremely lucrative contracts, and the Prime Minister, Kishi Nobusuke, was shown to be implicated with Kinoshita.[44] Again in 1959 the JSP quizzed the government about the role of Kubota Yutaka, the President of Nippon Kōei, in Vietnam reparations,[45] while more recently Den Hideo, Yokomichi Takahiro and Narasaki Yanosuke of the JSP pursued the controversy surrounding the development of natural gas (LNG) in Indonesia, in particular the intervention by a senior MITI official and his minister, Tanaka Kakuei, in commercial negotiations.[46]

The Opposition was also critical at times of the government's by-passing of the Diet in aid expenditure. In July 1967, for instance, Kitayama Airō if the JSP attacked the government for granting $1 million to Indonesia from *yobihi*, emergency funds allocated in the budget and able to be drawn on at the discretion of Cabinet. Kitayama considered the amount too large to be left to administrative fiat and demanded that Article 85 of the Constitution (that granting the Diet power of authorisation of expenditure) be upheld. The government countered this argument by claiming that the use of *yobihi* was legal under Article 87 of the Constitution (that providing for a reserve fund) and Article 35 of the Finance Law.[47]

Few chances to embarrass the government in the Diet forum were passed up by Opposition members, especially in the wake of the 1976 Lockheed scandal, when it would seem that every public medium was used to drag up major and minor sins. Indonesian LNG was one such case, as were the debates in the House of Representatives Budget Committee on Japan-South Korean relations, which had surfaced earlier during Lockheed hearings. Masamori Seiji of the JCP was successful in bringing to light what appeared to be discrepancies in pricing on rolling-stock for the Seoul subway, which was being financed by OECF loans under bilateral agreement. While no firm proof of misdemeanour was revealed, the publicity was extremely unfavourable. In addition, some Opposition members sought to advance their own interests not by criticising aid policy, but by promoting government aid to assist private firms. A DSP leader, in particular, used his position to force a favourable decision on government assistance to a Japanese private company's development project in Papua New Guinea, a case which will be taken up again in Chapter 7.

In most of these situations, however, it was the individual member, and not the party or the faction, who provided the initiative and energy for Opposition and backbench contribution to the aid policy debate. Factions certainly had their part to play, notably in top-level decision-making involving issues wider than aid (the decision to establish JICA is a good example), but aid did not impinge on party power politics in the same way that domestic or ideological issues did.

The aid bureaucracy in Japan, then, was well defined in formal allocation of responsibilities, but complex in its processes. Organisational weaknesses and imbalances, however, meant that only a limited sense of unity and community bound its scattered parts, and that officials, not advisors or politicians, defined the boundaries of policy. No coordination mechanism spanned ministries or effectively drew together the strands of aid policy – grants, loans, technical and multi-lateral assistance – although one did function actively for loans policy, as we shall shortly explain.

Aid structures fitted most comfortably into government when closely integrated with other policy concerns of ministries and agencies. What constituted aid policy depended significantly on the changing fortunes of the aid sections within ministries, since the political world did not support an expanded aid bureaucracy or back up an independent aid policy. At the same time, aid's limited political and organisational impact resulted in an aid bureaucracy which lacked energy. At the political level, perceptions of aid were confused; no sense of political directions for aid, let alone policy directions, was apparent. This left aid open to cross-cutting political pressures. Furthermore, the ruling Liberal Democratic Party took a less assertive role in aid at home than in many other policy areas. The explanation lay partly in the nature of aid itself, but also in the highly routinised procedures for aid management.

Notes

1. John White, *The Politics of Foreign Aid* (London, Bodley Head, 1974), pp. 290-303.
2. Ibid., p. 51.
3. Judith Hart, *Aid and Liberation: A Socialist Study of Aid Policies* (London, Gollancz, 1973), p. 187.
4. Dudley Seers and Paul Streeten, 'Overseas Development Policies' in W. Beckerman (ed.), *The Labour Government's Economic Record* (London, Duckworth, 1972), p. 152.
5. Judith Tendler, *Inside Foreign Aid* (London, The Johns Hopkins University Press, 1975), Ch. 4.
6. Nakane Chie, *Tekiō no jōken: nihonteki renzoku no shikō (Criteria for adjustment: the Japanese continuum mentality)* (Tokyo, Kodansha, 1972), Part 2.

7. See, in particular, Edwin Dowdy, *Japanese Bureaucracy: its development and modernization* (Melbourne, Cheshire, 1973); Albert M. Craig, 'Functional and Dysfunctional Aspects of Government Bureaucracy' in Ezra F. Vogel (ed.), *Modern Japanese Organization and Decision-making* (Berkeley, University of California Press, 1975), pp. 3-32; and various articles in Tsuji Kiyoaki (ed.), *Gyōseigaku kōza* (*Studies in administration*) (Tokyo, Tōkyō daigaku shuppankai, 1976), vol. 2, *Gyōsei no rekishi* (*Administrative history*). A summary of the bureaucratic control literature is made by Chalmers Johnson in 'Japan: Who Governs? An Essay on Official Bureaucracy', *Journal of Japanese Studies*, vol. 2, no. 1 (August 1975), pp. 1-28.

8. For example, in the popular literature such as Fujiwara Hirotatsu, *Kanryō no kōzō* (*The structure of the bureaucracy*) (Tokyo, Kōdansha gendai shinsho, 1974), or Kanryō kikō kenkyūkai (ed.), *Ōkurashō zankoku monogatari* (Bureaucratic Organisation Study Group, *Horror stories of the Finance Ministry*) (Tokyo, Yell Books, 1976).

9. Made clear in Nagano Nobutoshi, *Gaimushō kenkyū* (*A study of the Foreign Ministry*) (Tokyo, Simul, 1975); Honda Yasuharu, *Nihon neo kanryōron* (*A study of Japan's new bureaucracy*) (Tokyo, Kōdansha, 1974), vol. 1, or Nihon no kanryō kenkyūkai (ed.), *Oyakunin sōjūhō* (Japanese Bureaucracy Study Group, *How to use bureaucrats*) (Tokyo, Nihon keizai shimbunsha, 1971).

10. The Banking Bureau had jurisdiction over the affairs of the Eximbank. The Financial Bureau was involved in determining the annual allocation to Eximbank and the OECF from the Trust Fund. More will be said on this in Chapter 5.

11. Honda, *Nihon neo kanryōron*, vol. 2, gives details of the rivalry between MITI, MOF and 'proper' EPA officials in personnel policy. Chalmers Johnson examines this in the context of rivalry between MITI and MOF in 'MITI and Japanese International Economic Policy' in Robert A. Scalapino (ed.), *The Foreign Policy of Modern Japan* (Berkeley, University of California Press, 1977), pp. 237-44.

12. Yoshino Bunroku, a former councillor in the Bureau, was in 1976 Deputy Vice-Minister (*shingikan*) in the ministry, Sawaki Masao (Director 1970-2) served as Director of the Asian Affairs Bureau, and Katori Yasue (Director 1974-5) served as Director of the Minister's Secretariat, but no Vice-Ministers rose from directorship of the Economic Cooperation Bureau. Nagano, *Gaimushō kenkyū*, analyses the country 'factions' in his Ch. 2.

13. Private communication from an officer of the Japanese Embassy, Canberra, January 1976.

14. See Nagano, *Gaimushō kenkyū*, and Fukui Haruhiro, 'Policy-making in the Japanese Foreign Ministry' in Scalapino (ed.), *The Foreign Policy of Modern Japan*.

15. The dichotomy is analysed by Nakane Chie in *Japanese Society* (Berkeley, University of California Press, 1970), Ch. 3, and *Tekio no jōken*, pp. 131-6. M.Y. Yoshino, 'Emerging Japanese Multinational Enterprises' in Vogel, *Modern Japanese Organization and Decision-making*, discusses it in business firms, and Honda, *Nihon neo kanryōron*, vol. 1, pp. 117-55, considers the relations between the MFA and its embassies.

16. Yashuhara Kazuo, *Ōkurashō* (*The Ministry of Finance*) (Tokyo, Kyōikusha, 1974), pp. 145-65, and Asahi shimbun keizaibu, *Keizai seisaku no butaiura* (*Economic policy: behind the scenes*), (Tokyo, Asahi shimbunsha, 1974), pp. 242-6.

17. Gyōsei kanri iinkai, *Kongo ni okeru gyōsei kaikaku no kadai to hōshin* (Administrative Management Committee, *Problems of and policy for future administrative reform*), 2 April 1975.

18. George Cunningham, *The Management of Aid Agencies: Donor structures and procedures for the administration of aid to developing countries* (London, Croom Helm, 1974), pp. 16-20.

19. Nihon no kanryō kenkyūkai (ed.), *Oyakunin sōjūjō*, is an analysis of the bureaucracy along these lines.

20. Michel Crozier, *The Bureaucratic Phenomenon* (Chicago, University of Chicago Press, 1964). Hugh Heclo and Aaron Wildavsky, in *The Private Government of Public Money* (Berkeley, University of California Press, 1974), pp. 9 and 69-74, regard this informal communication as the essence of coordination.

21. Craig in Vogel (ed.), *Modern Japanese Organization and Decision-making*, pp. 7-8, has an excellent quotation from an unnamed MFA source. The officer substantiated the point made above that 'the atmosphere of the office either in Tokyo or abroad, depends a great deal on the personality of the men, particularly of the chief'.

22. For details on career patterns, see Okabe Shirō, *Gyōsei kanri (Public administration)* (Tokyo, Yūhikaku, 1967), Ch. 7, and Watanabe Yasuo, 'Nihon no kōmuinsei' ('Japan's civil service system') in Tsuji (ed.), *Gyōseigaku kōza*, vol. 2, pp. 111-60, and 'Kōmuin no kyariya' ('Civil service careers'), in *Gyōseigaku kōza*, vol. 4, pp. 169-207. See also Nihon no kanryō kenkyūkai (ed.), *Oyakunin sōjūhō*, pp. 9-60. There were eight grades of civil service officer in each of the three examination categories. Grade 1 to 3 positions were filled mainly by Higher Examination officers. In the Higher Civil Service Examinations, there were 29 types of examination ranging from Law to Agricultural Chemistry and the like.

23. Although certain division directorships were allocated for specialist officers in some ministries. Watanabe has charts in his 'Kōmuin no kyariya', pp. 187-90. A good analysis of this subject is Richard P. Suttmeier, 'The Gikan Question in Japanese Government: Bureaucratic Curiosity or Institutional Failure', *Asian Survey*, vol. XVIII, no. 10 (October 1978), pp. 1047-66.

24. A point made strongly in *Oyakunin sōjūhō*, which was written as a type of handbook for those wishing to lobby the bureaucracy.

25. Fukui, 'Policy-making in the Japanese Foreign Ministry'. The small numbers of career entrants were shown by reference to separate ministry totals: in 1973 those who passed the Higher Civil Service Examinations for the MOF numbered 27, Ministry of Education 12, Ministry of Welfare 18, MITI 39, Ministry of Transport 49, EPA 9, National Police Agency 21. For MAF the numbers were large at 118, but comprised mainly those who passed exams in agricultural science, only 16 passing in law or economics. Successful candidates for the Higher Foreign Service Examinations in 1973 numbered 27. See Watanabe, 'Nihon no kōmuinsei', pp. 148-9.

26. Cunningham, *The Management of Aid Agencies*, p. 42.

27. Bruce Juddery, 'Indigestion ahead as department swallows aid agency', *Canberra Times*, 4 June 1976.

28. Interview, 24 May 1976.

29. Tendler, *Inside Foreign Aid*, Chs. II and III.

30. Another example was the normalisation of relations between Japan and China. See Fukui Haruhiro, 'Tanaka Goes to Peking: A Case Study in Foreign Policymaking' in T.J. Pempel (ed.), *Policymaking in Contemporary Japan* (Ithaca, Cornell University Press, 1977), pp. 60-102.

31. Ehud Harari, 'Japanese Politics of Advice in Comparative Perspective: A Framework for Analysis and a Case Study', *Public Policy*, vol. XXII, no. 4 (Fall 1974), pp. 537-77.

32. OECD, Development Assistance Directorate, *The Management of Development Assistance: A summary of DAC countries' current practices* (Paris,

21 August 1975), p. 7.

33. The literature is voluminous in Japanese but in English, see Park Yung Ho, 'The Government Advisory Commission System in Japan', *Journal of Comparative Administration* (February 1972), pp. 435-67, for a comprehensive review of the facts and figures. T.J. Pempel, in his 'The Bureaucratization of Policymaking in Postwar Japan', *American Journal of Political Science*, vol. XVIII, no. 4 (November 1974), pp. 647-64, analyses their place in policy-making, while Harari's 'Japanese Politics of Advice in Comparative Perspective' gives a broader theoretical analysis of their place in government. Two important Japanese works are Okabe Shiro, 'Seisaku kettei ni okeru shingikai no yakuwari to sekinin' ('The role and responsibilities of advisory councils in policy-making'), *Gyōsei kenkyū*, no. 7 (1969), pp. 1-19, and Ogita Tamotsu, 'Shingikai no jittai' ('Realities of advisory councils'), *Gyōsei kenkyū*, no. 7 (1969), pp. 21-71. Most of these studies emphasise the susceptibility of councils to bureaucratic control.

34. The desk was established on 31 March 1965. The Councillors' Office of the Prime Minister's Office was staffed mainly by officers (of about division director level) transferred from other ministries. It was recognised as a key section of the bureaucracy, so much so that ministries referred to the number of 'shares' (*kabu*) they had in it, according to the officers they had temporarily assigned there.

35. They were the MFA, MITI, MAFF and Ministries of Education, Health and Welfare, Transport, Labour, Construction, Posts and Telecommunications, as well as the Environment, Administrative Management and Science and Technology Agencies. See Naikaku sōri daijin kambō shingishitsu taigai keizai kyōryoku tantō jimushitsu, *Taigai keizai kyōryoku shingikai tōshin (kaihatsu tojōkoku ni taisuru gijutsu kyōryoku no kakujū kyōka no tame no shisaku ni tsuite) no jisshi jōkyō nado ni kansuru shiryō* (Prime Minister's Office, Prime Minister's Secretariat, Councillors' Office, Overseas Economic Cooperation Desk, *Materials on the implementation status of the report of the Advisory Council on Overseas Economic Cooperation (Policies for strengthening technical cooperation to developing countries)*), 24 January 1975.

36. The Japan Overseas Cooperation Volunteers (Nihon seinen kaigai kyōryokutai), an organisation similar to the US Peace Corps, was a successful and respected group. Survey work was the basis of loans policy and was given momentum by broader political and commercial considerations.

37. See *Nikkei*, 10 June 1975. For the text of the Cabinet decision, see Minato Tetsurō, *Nihon no ikiru michi: kokusai kyōryoku no kihon seisaku o mezashite (Japan's lifeline: towards basic policy for international cooperation)* (Tokyo, Kokusai kyōryoku kenkyūkai, 1975), pp. 163-4. Members included the Foreign Minister, Finance Minister, Minister for Agriculture and Forestry, Minister for International Trade and Industry, Chief Cabinet Secretary and Directors-General of the Prime Minister's Office and the Economic Planning Agency.

38. Cunningham, *The Management of Aid Agencies*, p. 331.

39. Fukui Haruhiro makes this point in his survey of work on Japanese policy-making, 'Studies in Policymaking' A Review of the Literature' in Pempel (ed.), *Policymaking in Contemporary Japan*, pp. 42-4.

40. See OECD, Development Assistance Directorate, *The Management of Development Assistance*, pp. 15-19, and Cunningham, *The Management of Aid Agencies*, Ch. 2.

41. They were the Foreign Affairs, Cabinet, Budget, Finance, Audit, Trade-Industry, Agriculture, Construction and Social-Labour Committees.

42. Robert E. Ward, *Japan's Political System*, 2nd edn. (Englewood Cliffs, Prentice-Hall, 1978), p. 161.

43. See the various party handbooks prepared for the 1976 General Election.

44. Nishihara Masashi, *The Japanese and Sukarno's Indonesia: Tokyo-Jakarta Relations 1951-1966* (Honolulu, The University Press of Hawaii, 1976).

45. *Asahi shimbun*, November-December 1959.

46. 'Indonejia LNG meguru giwaku no kōzō 1-2' ('The structure of suspicion surrounding Indonesia's LNG, 1-2'), *Kokusai keizai* (October 1976), pp. 28-33 and (November 1976), pp. 39-43.

47. *Nihon keizai shimbun*, 14 July 1967.

4 MINISTRIES AND THE POLICY PROCESS

We now look more closely at the interplay of people and organisations in the aid process, how interministry coordination, and its obverse, were features of aid policy-making, and how procedures and processes were intertwined in the actions of officials. On paper, the outline of the aid administration minimised overlapping responsibilities, but the practice of managing aid was very different. The system of 'equal partnership' in aid policy, which suggested a degree of equality between ministries, was also characterised by participants as one of 'constructive competition', where sectional rivalries within and between ministries were a prime motive force in the development of policy. Aid policy was built up on bureaucratic decisions and outcomes, ranging from determination of total aid volume goals to the type of funds committed to recipients and the terms on which they were provided.[1] The ministries (and, in particular, the 'Big Four' of foreign aid, the MFA, the MOF, MITI and the EPA) directly influenced aid policies. Different procedures governed each kind of aid, and the way they were interpreted and carried out affected the shape and content of policy.

Figure 4.1 illustrates aid flows from the Japanese Government to the less developed countries and to multilateral agencies. The four main categories of aid — capital grants, technical assistance, multilateral and bilateral loans — were all separate, with distinct patterns of administrative control. The chart shows clearly how the implementing agencies channelled funds to multilateral and bilateral projects and how home ministries controlled these flows. Consultants and private enterprise helped survey and finance projects, and Japanese missions in recipient countries transmitted information between Japan, recipients and project management teams in the developing country. MITI, the MOF and the MFA were the ministries most actively involved in the aid process.

Bilateral Capital Grants

Bilateral capital grants, like bilateral loans, were intended to promote the economic and social development of the developing countries but, unlike loans, repayment was waived (that is, their grant element was

118

100 per cent). Grants were aimed mainly at the social infrastructure sectors which fell outside the scope of the financing regulations of the Overseas Economic Cooperation Fund (OECF), the Export-Import Bank (Eximbank) and JICA, and therefore included housing, education, medical and research fields. Capital grants were first made in 1969, when the emphasis was on industrial infrastructure projects, particularly in Indochina, but were later extended to social infrastructure, agricultural and marine projects and emergency aid (quite apart, of course, from $127.7 million in food aid given since 1968). Between 1969 and the first half of 1977, 15.1 per cent of grants were allocated to agriculture, 12.8 per cent to industry, 26.1 per cent to social infrastructure, 33 per cent to emergency aid and 12 per cent to 'other' (including a large grant of 5,000 million yen to Mongolia in 1976).[2]

In principle, capital grants were made mainly to the lesser developed countries, consistent with the practice of the Development Assistance Committee (DAC), but the Japanese Government contributed grant aid even to relatively advanced LDCs where there was a demonstrated need for funds for social infrastructure development, and where loans would be inappropriate. These latter cases were treated on an individual basis, and were premised on the country (such as South Korea and Paraguay) having 'friendly relations' with Japan.[3] By 1977, 80.9 per cent of grants had been given to countries with *per capita* GNP of less than $265, although LLDCs received only 23 per cent. In 1977 grants amounted to 6.2 per cent of Japan's ODA. The rationale underlying grant aid policy, as explained by the MFA, was that it should: (a) be given to the poorest countries; (b) be directed to social development projects (with some programme aid, such as food production assistance beginning in 1977); and (c) be tied to technical cooperation programmes. At the same time, Japan's interests were not neglected, for it was realised that aid for fertiliser could, for example, assist Japan's depressed fertiliser industry and aid to fisheries projects could promote Japan's aims in the era of 200-mile economic zones.[4]

Grant aid was the responsibility of the MFA Economic Cooperation Bureau's Second Economic Cooperation Division, in 1978 a group of 16 officers. Until 1978 this division acted alone in planning and implementing grant aid, but from April of that year a division was created within JICA to assist in implementing parts of the grant programme. Before then, the Second Economic Cooperation Division bore a heavy administrative load as grant aid commitments rose. A performance gap, indicated by a large carry-over of funds from one budget year to the next, was perpetuated by a general disinterest outside the ministries

Figure 4.1: Economic Cooperation: Structure and Process

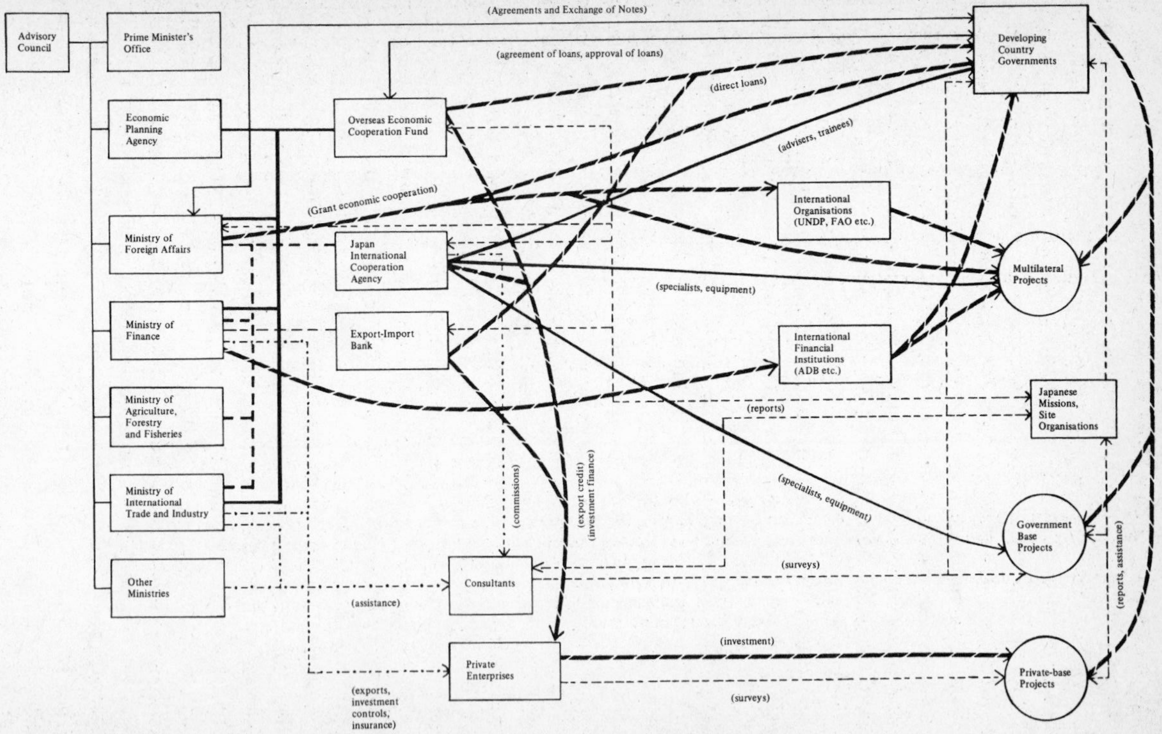

Key to Figure 4.1

1. Aid Flows

Capital flows, including:

— capital grants from the Ministry of Foreign Affairs;
— multilateral aid from the Ministry of Foreign Affairs to international organisations (UNDP, FAO etc.);
— multilateral aid from the Ministry of Finance to international financial institutions (ADB etc.);
— funds from the Overseas Economic Cooperation Fund in the form of direct loans to developing-country governments (grant element 25 per cent +) and investment finance to private enterprises;
— funds from the Export-Import Bank in the form of export credits, investment finance and loans (grant element up to 25 per cent);
— funds from the Japan International Cooperation Agency in the form of low-cost investment finance to private enterprises.

Technical aid flows, to and from the Japan International Cooperation Agency, including:

— advisers to developing-country governments;
— specialists and equipment to multilateral and bilateral projects;
— trainees from developing countries.

2. Administrative Controls

Over the Overseas Economic Cooperation Fund by the Economic Planning Agency, Ministry of Foreign Affairs, Ministry of International Trade and Industry and the Ministry of Finance (the four-ministry loans committee).
Over the Japan International Cooperating Agency by the Ministry of Agriculture, Forestry and Fisheries, the Ministry of International Trade and Industry, the Ministry of Foreign Affairs and the Ministry of Finance.

By:

— the Ministry of International Trade and Industry and the Ministry of Finance over private enterprises, in the form of export and investment controls, export insurance etc.;
— the Japan International Cooperation Agency (and ministries) over consultants, in the form of commissions for consultations.

3. Information

Consisting of:

— surveys by consultants and private enterprises and survey reports to developing-country governments and to Japanese embassies (and then to JICA);
— reports to and from Japanese Embassies in developing countries to home agencies and to the Ministry of Foreign Affairs. Also communication between embassies and the developing-country government.

4. Official Contacts

Between:

— the Japanese Government and the developing-country government as exchanges of notes and government-to-government agreements;
— the developing-country government and the Overseas Economic Cooperation Fund as approval of loans and loan agreements.

Source: Administrative Management Agency documents.

in grant aid, which went to projects small, inexpensive and unspecta-cular in comparison to loan-funded projects. Furthermore, the need both to approve and implement the grant programme meant more desk work for the division's officers. The 1977-8 pronouncements on doubling Japan's ODA threw a greater onus onto the grant aid division to speed up disbursement and simultaneously handle a greatly expanded budget (increased from nearly 16,000 million yen in fiscal 1975 to 39,000 million yen in fiscal 1978).

The new JICA division was designed to assist in this, by taking over the implementation of grants from the exchange of notes stage of bilateral negotiation until actual payment. JICA's involvement was justified on the grounds that most grant cases (25 out of the 26 budgeted for in 1977[5]) had some connection with technical assistance, but emergency, cultural, food and food-production aid were, as before, managed by the MFA division (the last two were budgeted as an MOF allocation and requested by the MOF's International Finance Bureau). In fact, it was estimated that over one-third of the items in the 1978 grant budget would remain fully in MFA hands and would not be implemented by JICA. The potential to achieve a better mix of tech-nical and capital grants work was institutionalised, although the need to control the grant programme became more acute.

Grant assistance was initiated by a request from the developing country, followed (if necessary) by a study of the feasibility of the project, an MFA request to the MOF for an allocation in the following year's budget, an exchange of notes with the government of the developing country and, finally, the arrangement of banking, settlement of contracts, issuance of export licences and payment of moneys from general revenue into the recipient's account in an authorised foreign exchange bank. The process was unlike that for loans in several ways, but particularly in one respect: budget allocation had to be secured before the exchange of notes between governments took place. Or, more formally, 'in principle, implementation must follow the exchange of notes and be completed within the single-year budgeting frame-work'.[6] In one way, this gave the MFA the advantage of implementing grants immediately after making a commitment, but it also implied that until all budget formalities were finalised, no commitment could be entered into by the government.

Grant aid was subject to bureaucratic rivalries even when the poten-tial recipient first approached the Japanese Government. There was no multiministry system of control of grants as there was of loans or technical cooperation. The decision on whether requests would be acceded to or not lay with the Second Economic Cooperation Division and with other sections of the MFA and also with the MOF Budget Bureau via the International Finance Bureau's Overseas Investment Division. Since commitments of grants depended on budget appropriation, it was important to have understandings with the MOF at an early stage, certainly before an official budget application was made.

The Economic Cooperation Bureau's Policy Division was the first to judge requests for aid received by the Japanese Government. It formed its own opinion about the type of aid most suitable (if unspecified) and in passing the proposal along to the specialist division, weighed the request in the light of ministry policy and precedent. The Second Economic Cooperation Division could refuse a request on its own initiative, which it sometimes did,[7] although the final arbiter was the Bureau Director. The division made an appraisal after wide consultation, but favoured requests

1. which were 'mature' and for projects likely to be completed;
2. for projects which had priority in the recipient's own economic plans;
3. from those countries which could 'genuinely be assisted' by Japanese grants;
4. for projects which would not become associated with commercial profit;
5. for projects which had strong links with technical cooperation.

These criteria did not necessarily match those of the MOF or of other sections of the MFA, nor were they sufficiently precise to prevent ambiguity in assessing 'appropriateness'.

Regional bureaus in the MFA put their opinion about the advisability of a grant to the country in question, based on their assessment of its wealth, strategic position, relations with Japan and so on. Some Economic Cooperation Bureau officials asserted that regional officers did not consider need or feasibility but made purely political judgments,[8] so there was a distinct divergence of approach, although in most instances it was not wide. Regional bureaus were involved less with grants than with loans, which had greater political and diplomatic relevance (because of their visibility and monetary value), and therefore these bureaus followed grants policy less vigorously. The country distribution of Japan's grant aid, however, especially the bias in the early years towards South Vietnam, was obviously related to political considerations beyond the Second Economic Cooperation Division's five guidelines.[9]

Once a decision was made in the MFA Second Economic Cooperation Division about the advisability of the grant, budget funds had to be secured. A request was made, as part of the annual MFA budget application, for an amount expected to be agreed upon in exchanges of notes in the coming fiscal year. Official discussions with the MOF took place both at the time of the original budget request and when the MOF examined the state of the project and the need for disbursement. Since the number of grants made each year was relatively small,[10] the Budget Bureau had ample opportunity to scrutinise each one. The MOF was wary of capital grants since gratuitous aid was not, in its opinion, economically sound. Although the grant budget increased substantially in 1978 to assist aid-doubling goals, MFA demands in 1975-6 for increased grant aid to balance the fall in reparations had met with MOF insistence on more efficient use of grants, greater LDC efforts to help themselves and continued high visibility for grant projects.[11] The MOF seemed to assume that visible grant projects somehow offset the financial burden on Japan. Low grant budgets had always made difficult a wide geographical spread of aid; a policy on geographical distribution was hard to lay down. Until 1975 most grants had been given to Indochina, notably South Vietnam. The only four countries outside Asia to receive grants up to 1975 were Tanzania, the Maldive Islands, Guyana and Papua New Guinea, but 1976-7 saw a sudden flowering of grant aid and ten new non-Asian recipients were included.[12]

This routine was cumbersome and staff shortages in the MFA Second Economic Cooperation Division meant that it worked slowly once budget approval was given; JICA's help was intended to speed up the process. Negotiations with the recipient government on the exchange of notes were time-consuming, for in principle only when the budget was passed could talks begin and the Second Economic Cooperation Division therefore had to conduct them within a strict budget framework. Any alterations to the budget for individual projects needed MOF imprimatur. The requirement to negotiate and make commitments within a precise budget limit removed all flexibility from a system already under strain. From April 1978 the Second Economic Cooperation Division managed all stages of implementation of the grant up to the exchange of notes, and whereas it had previously supervised and verified the letting of tenders and contracts with Japanese suppliers by the recipient government, this was taken over by JICA from 1978. The MFA co-ordinated payments and remained intimately involved in the management of projects, in order to watch the disbursement of funds.

Despite the rule that projects unable to be completed in the same fiscal year as an exchange of notes was concluded should not be accepted, grants often took years to complete. The low rate of disbursement of grant budgets was evidence of that fact that grant funds were not easily spent in the same fiscal year in which they were allocated.[13] Until the political initiative of 1978 in the form of promises to expand Japan's whole ODA programme, the grant aid administration was caught in a situation which prevented aid being expanded to counter criticism of Japan's lagging grant performance. The MOF was in principle opposed to any large increase in grant aid from the budget unless sound agreement on policy for grants could be reached, but the nature of grant aid was such that only temporary agreement between the MOF, the Second Economic Cooperation Division and other sections of the MFA was likely. Grant aid was not the subject of debate at a higher policy level in the now defunct Ministerial Committee or in the LDP Special Committee. The Advisory Council did recommend the expansion of grants, but could not enforce its proposals. While budgets remained low, however, any setting of guidelines was impossible, for a geographically and sectorally balanced array of projects could not be planned. Officials explained that previous policy was maintained for want of funds.

This extremely restricted scenario for grant aid changed to some extent in 1978, and MFA officials faced the challenge of both expanding and speeding up the programme. They had also to confront several policy questions, quite apart from overseeing JICA's implementation. A bigger budget was likely to bring demands from two sides — from one to lay down more exact guidelines on the use of grant allocations, and more flexible provisions of local cost financing and untying, and from the other to align grants with Japanese diplomatic, political and fiscal interests. It was likely that the MFA's Second Economic Cooperation Division would be drawn into more debate with MFA regional bureaus and other ministries. The tendency for implementing agencies to develop in time their own influence over future policy options (as will be explained in Part III) was possible also in grants, although in 1979 the MFA retained control of how the grant budget would be spent.

Multilateral Aid

There were two main flows of multilateral funds: grants from the MFA (and from other ministries) to the United Nations and its related organisations; and loans and grants from the MOF to international financial institutions such as regional development banks and the World Bank group. They were two distinct policy areas managed by separate administrative units. As pointed out in earlier chapters, Japan was always very positive in its attitude to multilateral aid and maintained a record of contribution to multilateral institutions somewhat better than the DAC average (in 1977 Japan's multilateral assistance as a percentage of ODA was 36.9 per cent as against a DAC average of 31.4 per cent). This was due in no small part to Japan's embracing the United Nations as a basis of its postwar foreign policy and its willing participation after the mid-1960s in the GATT-based international economic system. Multilateral aid had direct benefits for policy-makers, according to the MFA: in contrast to the overt foreign policy uses of bilateral aid, multilateral contributions were politically neutral and were able to be put to good use by the aid specialists in multilateral agencies.[14] As will be shown, however, the loss of Japan's donor identity in the eyes of recipients of multilateral aid was certainly a worry for officials, although perhaps somewhat offset by the kudos bestowed on Japan by its fellow donors.

Japan contributed from its budget to six international financial institutions: the International Bank for Reconstruction and Development (the World Bank), the International Development Association (IDA), the International Finance Corporation (IFC), the Asian Development Bank (ADB), including its Special Fund and Special Technical Fund, the Inter-American Development Bank (IDB) and the African Development Fund (AfDF). In addition, some 30 United Nations-sponsored programmes received Japanese support, as did other international bodies mainly in Asia, including the Asian Productivity Organisation, the Southeast Asian Fisheries Development Centre, the International Fund for Agricultural Development (IFAD), the Red Cross and others. Between 1971 and 1977 both forms of multilateral aid increased significantly, from $11.1 million donated to UN and other agencies and $58.5 million to international financing institutions, to $74.3 million and $456.7 million respectively.

Multilateral assistance consisted of more than the carrying out of obligations determined by international agreements. It was true that the outlines of multilateral aid policy were set by the broad compass of Japan's role in these organisations but translating that involvement into actual policy drew in a number of ministries and their divisions. The MOF was represented by the Overseas Investment Division of the International Finance Bureau and the MFA by the United Nations (UN) Bureau and by the International Organisations Division of the Economic

Cooperation Bureau. Because many contributions to multilateral organisations were generally governed by international treaties and agreements, they were automatically included in the budget, but their size was determined within the budget in consultation with the MOF's Budget Bureau. Debate on multilateral aid policy, therefore, was restricted to a narrow group of officials concerned with the disbursement of that assistance. The coordination of separate flows was effected mainly by budgeting (and, for some items, ministerial level decision), since no part of the aid administration had the jurisidiction to monitor all multilateral aid.

The conditions of Japan's membership of international financial institutions were set by the charters of those bodies and by domestic legislation enabling Japan to take part. The National Diet approved bills to alter any aspect of Japan's role and the timing of Japan's contributions fell within a broad legislative framework. Payment from government funds or in government bonds was commonly on a three-yearly basis, with annual instalments, which meant that it was necessary to renegotiate with the Budget Bureau at three-yearly intervals instead of annually, although yearly requests for instalments still had to be made by the International Finance Bureau. Non-regular payments, such as those to the Asian Development Bank Special Fund, were requested as required.

There was no sustained interchange between the MOF and the MFA on the size and scope of multilateral aid budgets, although they did consult (especially on new requests), and it would be fair to say that, in view of the generally favourable response to Japan's effort, both ministries agreed on the need to maintain a healthy multilateral aid programme. None the less, there were complications: the International Finance Bureau was in a sensitive position as the MOF's 'window' for aid, because it had to argue in favour of aid to the Budget Bureau while at the same time tempering its more independent view to correspond with stricter MOF attitudes on budget expenditure, an outlook spurred by the dominance of the Budget Bureau and the defensive, domestic orientation of the bulk of the ministry.[15] Relations between the International Finance and Budget Bureaus were not always easy, given that the former recognised the importance of Japan's international behaviour to domestic economic prosperity, a perception which the Budget Bureau shared less fervently. Differences in attitudes such as this fostered the International Finance Bureau's 'foreign' image within the MOF, although its aid requests were not treated severely by the Budget Bureau, in whose catalogue of tests 'fiscal soundness' was always uppermost and to whom overseas criticism of Japan's aid procedures was of lesser moment. This intraministry conflict was not unusual, but the form it took — a curious 'double advocate' role — made life difficult for officers of the International Finance Bureau. While acknowledging this, they were still part of the MOF and committed to the effective use of Japan's multilateral assistance. The bureau was not reluctant to express itself on this subject, as it did in 1976 in the argument about replenishment of the International Development Association, or indirectly in the comments on the World Bank's aims and management made by a retiring Bank Director and former MOF official, Hori Tarō, in 1976.[16]

The International Organisations Division of the Economic Cooperation Bureau was the MFA's counterpart of the Overseas Investment Division of the MOF's International Finance Bureau, and its brief covered 'international organisations and conferences', which made it the point of contact with the DAC.[17] The division assisted in the development of ministry policy on multilateral aid and many of its views overlapped those of the International Finance Bureau of the MOF. Both saw a need for balance between bilateral and multilateral assistance (which implied support for the existing level of multilateral aid) and for continued emphasis on regional banks. The close liaison between the International Finance Bureau and the Asian Development Bank since the bank's inception was one reason for the MOF favouring regional financing, but a desire first for the visibility of aid, and secondly for its efficient management, was also to the forefront of MOF thinking. These criteria were valued too by the MFA. The International Organisations Division acknowledged the trend to regional organisations but suggested at the same time that aid to the African Development Fund involved a loss of recognition of Japan in Africa, which did not occur, in contrast, with aid to the Asian Development Bank; multilateralism had some limits.

Japan's relations with the bank, and the way in which the bank's activities were related to Japanese bilateral policies, are matters of some dispute. There is no denying Japan's formative role in the creation of the bank in the years leading up to 1966. One writer even credited former Prime Minister Kishi with some influence, because of the Asian Development Fund he proposed in the late 1950s and which was later used to found the OECF.[18] Japan has always strongly supported the concept of an Asian regional financing institution, some would say for conscious political, others would suggest economic, reasons.[19] John White, in an early assessment, observed the United States-Japan axis of the bank's operations, but noted at the same time that there were strong developing-country forces ranged against this domination.[20] Other writers have disputed White's claim that the bank was not Asian, although most have agreed on the significance of developed-country interests and perceptions. Haas argued that Japan and the US were the main weight 'in setting the tone' of deliberations and that the US 'defers to Japan in matters of policy'. Japan was, he wrote, 'the country that has assumed a leadership role'.[21]

The President of the Bank has always been Japanese — Watanabe Takeshi from 1966 to 1972, Inoue Shirō 1972-6, and from 1976 Yoshida Tarōichi. Both Watanabe and Yoshida were formerly in the MOF, while Inoue spent his career in the Bank of Japan. Successive Japanese directors have been senior MOF officials, and many Japanese are on the staff at any time. Japan in 1977 provided 20.4 per cent of the bank's capital and held 16.7 per cent of the voting rights (22.8 per cent of regional votes). These facts, however, do not of themselves prove Japanese dominance of bank policy; they do indicate, nevertheless, the considerable scope for interplay between Japanese Government interests and bank operations. How this affected lending policies depended on the factors in individual cases, but the benefits of an obvious Japanese presence were well recognised by Japanese officials; Japan in this sense was very consciously an Asian nation.

The MFA International Organisations Division had only a small budget of its own to implement, much of which was paid to the Asian Productivity Organisation. The large amount of aid granted to United Nations agencies was the responsibility of the UN Bureau, and the Economic Cooperation Bureau communicated officially with that bureau through the Policy Division. Assistance to United Nations programmes was given either under multiyear or annual agreements, and the level of contribution was decided within the budget framework. The MOF, however, regarded the United Nations aid as 'rather too political' (as a result of the increasing politicisation of North-South issues in the UN forum itself and individual UN programmes) and thought the international financial institutions to be the 'strongest' and, therefore, the most appropriate channels for Japan's multilateral aid. Allocative efficiency was seen to be maximised in giving to the specialist lending agencies. In addition, the strained relations between the MFA's UN and Economic Affairs Bureaus complicated the development of policy towards United Nations aid programmes. Rather predictably, officials at the working level accused each other of taking an unbalanced view of the United Nations and UNCTAD, and even of administrative incompetence. Both bureaus had jurisdiction in

UNCTAD matters and they were in some conflict over policy towards UNCTAD IV in early 1976. The main controversy (in simple terms) was on the question of the Common Fund, which the Economic Affairs Bureau opposed in defence of the free market system. That bureau was regarded as representing the position of the advanced nations, in opposition to the UN Bureau's proximity to LDC arguments. The Economic Affairs Bureau was also at odds with the Economic Cooperation Bureau policy towards the Conference on International Economic Cooperation (CIEC) in Paris, for which the Economic Affairs Bureau had the main responsibility. Disagreement centred especially on the debt question. One Economic Cooperation Bureau official stated that 'we do not see eye-to-eye at all with the Economic Affairs Bureau', which he considered had no understanding of the problems of the North-South issue.[22]

The third aspect of multilateral aid policy was the management of relations with the DAC. In Part I it was explained how Japan joined the DAG and then the DAC, and how at the time she saw the opportunity as contributing as much to her acceptance in the international economic community as to her aid policy. There was little agreement between officials, however, on the impact of the DAC on Japanese policy over the years. To some, the DAC was an unknown quantity (even an 'outcast area' (*tokushu buraku*)[23]) until 1970, when the DAC High-Level Meeting was first held in Tokyo. This meeting, at which agreement was reached on the removal of tying from aid, demonstrated to other ministries the DAC's force as a donor organisation. It helped to raise consciousness of aid in Japan.

Membership was something of a status symbol for Japan in the 1960s and the MFA took pride in even this limited recognition of Japan's aid role. The MFA did not accept DAC views wholeheartedly — a Director of the Economic Cooperation Bureau has referred to the DAC rather disparagingly as 'a group of idealists'[24] — but its attitude towards the DAC was far more positive than the marginal interest shown by the MOF, although both took membership seriously. They acknowledged that the DAC affected Japanese perceptions of untying, terms, volumes and sectoral aid programmes, but MOF officials were less inclined to agree that DAC influence extended to policy as implemented, or to the details of policy: they noted what the DAC said and accepted the need for a common aid effort, but, in their opinion, procedures remained a domestic preserve.

The DAC forum was, as Japanese officials had foreseen in the early 1960s, an essential entrée into, and window on, the international aid debate. Japan's objectives in entering the DAC were well secured, but her obligations as a member extended to participating in the debate. Japan's influence within the DAC was restricted, however, mainly because of her slowness to soften her aid terms and her historically poor performance. She was not at the centre of DAC policy-making or DAC promotion, unlike other large donors, especially the United States. The MOF apparently preferred the country to take a low profile, wishing to isolate Japan (and, by implication, her aid budget) from the incipient trend to a greater political role for the DAC in the North-South debate. Japan's passive stance over the years served to insulate the bulk of the Japanese aid administration from the direct impact of DAC initiatives and deliberations. Language was another problem, and was apparent not so much in the Japanese presence at DAC meetings, but in the slowness with which information filtered through the system. There was always a lag because of the need to translate material before it could be widely used by the bureaucracy. The argument that 'the Japanese were one year behind' in their support of new themes, while somewhat exaggerated, does give a sense of the difficulties faced by Japanese officials in keeping abreast of trends and in ensuring that they were reflected in Japan's aid policy. The MFA was the important articulator of DAC policy developments — its support for the Basic Human Needs Programme is a good example — and it helped to marshal what were often positive Japanese responses at subsequent DAC meetings.

The DAC could not, however, be entirely insulated from the domestic politics of aid: the impact of external arguments on internal processes were substantial in this case. The MFA regarded the forum as a useful adjunct to its own relatively liberal aid position, although it was selective in referring to DAC statements for support, and it sought to use DAC censure of Japan's aid performance in its own power politics, particularly at budget time. It was also keen to expose other ministries to criticisms of Japan raised in the DAC and to 'work on' International Finance Bureau officials freed of the apron-strings of the Budget Bureau.[25] International Finance Bureau reports to its own ministry on the DAC examination were important in maintaining the thrust of this exercise. Nevertheless, not all officials accepted MFA arguments on aid and sometimes used the DAC session against the diplomats: representatives of one of the Japanese implementing agencies at a recent DAC Examination reportedly lobbied United States representatives to put pressure on the Japanese delegation, especially MFA officials, over its aid policies.[26]

Contact with the DAC was managed through the International Organisations Division of the MFA, and the Japanese team which visited Paris for the annual examination included officials from this division, the International Finance Bureau of the MOF and the Economic Cooperation Department of MITI. In preparation for the annual examination, the government compiled in mid-year a memorandum on its achievements of the previous year. The International Organisations Division of the MFA prepared the first draft, which was then circulated to other ministries for comment. Preparing the memorandum was not complicated — officials said that a harder task was the drafting of replies on the list of questions sent by the DAC Secretariat as a basis for the examination — although wording naturally indicated subtle differences between ministry attitudes, more so because of the need to translate the memorandum into English. While the MFA line tended to dominate (especially in translation), MOF caution tempered what its officers often regarded as overly optimistic or expansive statements.

The MOF's International Finance Bureau, however, again had to manage internal sectional conflicts over the bureau's position on the draft. Its Overseas Investment Division had to consult with the Budget Bureau on the fiscal implications of the draft, and with the Overseas Public Investment Division. This latter division was concerned exclusively with managing government loans and tended to regard its neighbouring division as amateurs on bilateral loans, discussion of which took up a large part of the memorandum. Accordingly, Public Investment put its view forcefully to the Overseas Investment Division (the regular channel of communication with the MFA on the memorandum and, in administrative terms, a more powerful division within the bureau), to check its grasp of loan details. The Overseas Public Investment Division's lack of confidence in the ability of the Overseas Investment Division to effectively argue for the MOF on loans led it as the specialist division to try to bypass regular procedures and talk directly with the MFA. Relations of trust within ministries, it would seem, were necessary for successful interministry negotiation.

Technical Assistance

Technical assistance mainly involved flows of human resources — people and their many skills — and was the most diverse form of aid given by Japan. The administrative arrangements for the development and management of policy were extremely complex. Figure 4.1 shows flows of technical assistance emanating from JICA, an agency which was controlled by the MFA in conjunction with the MAFF, MITI and the MOF. These flows included the movement of equipment, Japanese advisers and specialists, Japan's intake of developing country trainees and the conducting of feasibility surveys, often through Japanese and overseas consultant engineering firms. Other minor technical aid was conducted (or commissioned of JICA) by several specialist ministries, including Construction, Health and Welfare, Posts and Telecommunications and Education, and by local governments.[27]

For many years the DAC criticised Japanese technical aid efforts which, by comparison with other large donors, were meagre. Japan's annual disbursements of technical assistance certainly rose over the 1970s, jumping from $27.7 million in 1971 to $221.2 million in 1978 but when set against other donors Japan's performance was weak, ranking 12th out of 17 in the DAC in 1978, with just 9.9 per cent of ODA being given in technical aid. France topped the list with 51.4 per cent, followed by Belgium (35.5 per cent) and West Germany (34.3 per cent); the United States ranked lower than Japan at 15th with 7.5 per cent.[28] This performance gap was long recognised by Japanese officials but intractable problems of policy-making and implementation, coupled with a preoccupation with bilateral loans, prevented real improvement. Technical assistance, being another form of grant aid, was subject to Budget Bureau influence over the annual allocation, which was one cause of restraint, but weak coordination prolonged the struggle of sections of the technical aid administration to expand their programmes. In response to the Advisory Council's 1971 report on technical cooperation mentioned in Chapter 3, for example, some smaller programmes were upgraded but the main components of technical aid policy were left much as before.

Although the technical aid budget was divided between a dozen ministries and agencies, only four ministries were closely involved in technical aid policy-making. The MFA's Economic Cooperation Bureau contained three divisions which carried the bulk of technical aid management, while MITI and the MAFF had specialised technical aid divisions or desks. The MOF participated in its budgeting capacity but the EPA and the MOF's International Finance Bureau had no jurisdiction. In principle, the system was of a different character to those encountered in capital grant or multilateral aid, not simply because the participants were different, but because an outside agency — JICA — implemented most policy. While no permanent grouping of ministries existed to supervise technical aid, however, the dependence of JICA on the specialist ministries for staff to man its technical aid programmes threw much implementing of policy onto the policy-*making* machinery. Combined with different technical assistance programmes carried out by a dozen or so government bodies, this led to some confusion and prevented parts of the system from cooperating effectively. When there were six main kinds of technical aid requiring separate request and approval procedures — trainees, despatch of experts, survey missions, overseas volunteers, technical projects and trainee schemes run by 26 prefectures — it was important that an overview was maintained.

The Economic Cooperation Bureau of the MFA boasted three technical cooperation divisions, relations between which made managing JICA an unwieldy exercise:

1. The First Technical Cooperation Division handled technical assistance provided independently of projects, such as the despatch of individual experts in response to isolated requests. The burden of JICA supervision fell mainly to this division, which also undertook technical aid planning.
2. The Second Technical Cooperation Division was in charge of 'project base' technical cooperation, or the supply of materials and personnel for technical aid to particular projects, especially training centres. The Second Division separated from the First in 1972 in response to increased budgets for technical assistance and the resulting pressure on the resources of the existing division. The demarcation made was not altogether successful, for it imposed a false barrier between roles which often overlapped. The break was, however, in keeping with the functional organisation of the bureau.
3. The Development Cooperation Division was established in 1975 to assist the coordination of official and private technical cooperation, which specifically included the supervision of JICA development surveys.

The MFA divisions were the focal point of the technical aid process, for they handled MFA's dominant role in JICA management, received all requests for technical assistance and, of any of the participating ministries, had the most secure vantage point over the field of technical aid. The MFA shared responsibility for specialised tasks with other ministries, and in fact relied on them for providing much of the expertise and facilities for implementing policy. Mining, industry and energy, for example, fell under MITI jurisdiction. The work of the Technical Cooperation Division of MITI's Economic Cooperation Department was divided into two main aspects, personnel and projects. The division administered the training of developing-country students, the despatch of specialists from Japan and the organisation of training centres in recipient countries. Project assistance involved the planning and surveying of potential development projects by the Mining and Industry Departments of JICA.

The MFA and MITI Divisions stood aloof in technical cooperation, for there was little overlap of formal authority. The MFA presumed that it was dominant and some officials tended, rather unrealistically, to regard MITI as a subordinate appendage with little power over JICA. They saw no need for discussions with MITI about the agency. These attitudes were, however, products of a serious MFA weakness — it could not provide any technical specialists of its own — and reflected an aggressive defence of MFA authority in a policy area where much informal influence rested not only with MITI, but also with the MAFF, Construction and other technically oriented ministries which provided JICA with most of its specialist manpower for exchange schemes. MFA attitudes neglected the importance of mining, industry and other sector technical aid to the Japanese programme and the strong emphasis on the mining and industry portion of feasibility survey work.[29]

The technical aid process (for all types) supposedly began with a request for assistance from the developing-country government to the MFA through the local Japanese diplomatic mission. This was a procedural, not a legal, requirement, but requests were not necessarily inspired by the LDC government.[30] The stated policy of the Japanese Government was a passive one, whereby aid could be initiated only by request but, as Cunningham shows, for Western donors the balance of the active and reactive in a government's aid role could vary between donors and between parts of a donor administration.[31] While the Japanese did not recognise the value of an active approach to aid-giving, their faithfully maintaining the appearance of non-intervention in recipient policies towards aid often had ridiculous consequences.[32] For the LDCs, however, the result could also be extremely serious, even damaging to developing-country priorities. The overriding Japanese concern to avoid any stigma associated with being an active donor was overcome only when officials relied on the vigour of non-official participants in aid relations to compensate for opportunities lost through official caution. This encouraged the proliferation of technical missions which the LDC neither wanted nor could control, the well-known phenomenon of 'survey pollution' (*chōsa kōgai*).

Once a request was received by the MFA, a decision had to be made on whether to go ahead with the proposal. Several factors entered here, for while the MFA had the right to refuse, it usually consulted with JICA and with other ministries on details, and the extent of that consultation was determined by the MFA. In principle it did not pass on information about every request to other ministries or even to JICA, to prevent unnecessary interference (especially by MITI), but interministry rivalries usefully asserted themselves: the informal communications network operating through ministry representatives in overseas embassies or JICA offices ensured effective flows of information outside MFA channels. As one MITI official put it, 'We would find out in any case, so there's no point in the MFA refusing a request without checking with us.' In theory, properly coordinated thinking on policy would require a flexible MFA approach to this question, but reality was less than perfect.

As the MFA had no monopoly on information, so its capacity for technical appraisal was limited and it depended on, for example, MITI to assess proposals in mining and industry sectors. MFA authority was exerted over broader policy and political issues. Unacceptable requests were usually obvious, particularly those which lacked sufficient detail or those which would clearly breach Japanese policy if accepted, such as requests from South Korea for assistance with nuclear technology. On mining and industry requests, MITI decided whether to approve the aid or not, and the MFA could not force MITI to support a proposal which it considered impracticable or undesirable. The same principle applied, by and large, to those requests requiring cooperation from other specialist ministries. There was, in a real sense, an active balancing of interests between the ministries, despite the nominal supremacy of the MFA in policy planning and JICA supervision.

The acceptance of requests did not necessarily imply the approval of aid requested although, except in the case of large and complex requests, it did suggest tacit commitment. Proposals had to be checked against the availability of funds and, importantly, of personnel, and against the JICA budget limits which were fairly restricted. There also had to be indication of whether or not specialists and/or equipment could be allocated. As with grant aid, the scope of Japan's technical assistance was defined to a large extent by the budget. Quantity was naturally affected, although this depended a great deal on the nature of ministry budget requests and on their appraisal by the MOF. On the other hand, regional distribution was more directly influenced by decisions made in the MFA and MITI, often as a result of exchanges with regional divisions but nevertheless still as a part of preparations for budget requests. Asian countries have naturally benefited most from this distribution, although in the late 1970s there appeared to be some convergence towards a ratio in the rough proportions of Asia 5, Middle East 2, Central and South America 2 and Africa 1. This could not be maintained in all aspects of technical aid (given the diversity of procedures) but was approached in those areas most susceptible to MFA control, trainees and specialists.[33]

Certainly budget allocation was necessary in principle for the implementation of aid proposals, although budgets were not decided for each case individually (in contrast to capital grants) but rather for total numbers of types of specialists or quantities of materials. There was considerable flexibility within the JICA budget and much discretion left with the MFA and MITI in the division of the total budget into specific areas or projects selected from the backlog of aid requests. Firm decisions about the actual aid to be undertaken during the financial year were made between December and March, using as a guide the list of possible projects on which the budget was requested. At that stage some exchange took place between the Technical Cooperation Divisions and the Policy Division of the Economic Cooperation Bureau and between regional bureaus of the MFA. The necessary approaches were also made to other ministries about the use of technical personnel for chosen projects. Final explanations were made to the Budget Bureau at the start of the fiscal year.

While the budget was a strong constraint on decisions affecting the content of policy, other structural problems were more immediate, notably those relating to the supply of qualified personnel. This was commonly cited as the chief barrier to improved technical cooperation, both by officials and by other observers, but few substantial reforms were made, apart from the gradual betterment of salaries and conditions, registration, pooling and so on. MITI made in 1978 what appeared to be the most positive step forward, in seeking budget funds to mobilise specialists in private enterprise for work in LDCs, and secured a sevenfold increase in its budget allocation for this in 1979. The system required JICA to apply to specialist ministries or, more commonly, private enterprise for personnel to be drawn from their own staff or from the reserve in subsidiary organisations. Likewise, JICA had to request ministries to accept overseas trainees in their training programmes. It was an *ad hoc* and unsystematic arrangement which would not ensure that in the long term adequate and suitable personnel would be available.

Technical assistance policy depended primarily on short-term factors and on the extent of cooperation between ministries for the use of staff. No forward planning capacity was evident in the technical aid administration; no division sufficiently controlled resources to be able to plan beyond the single-year budget, although indicative yearly plans for the regional and country distribution of technical assistance were drawn up to prepare budget requests (see Chapter 5). JICA was a large and composite agency whose scope for policy initiative independent of the ministries was narrow.

Government Bilateral Loans

Yen loans were the core of Japanese foreign aid policy and dominated official thinking about aid since Japan first became a donor. Loans gave the most pressing impetus to official aid, for through them Japan was tied politically and commercially to the world's developing nations — as, in debt, were they to Japan. Loan aid was an essential element of the country's foreign and economic policies and its roots did not lie in the budgeting process or in the inadequacies of the administrative system, but in the broad reaches of Japan's relations with the developing countries. Loans were not merely a problem for governments; they attracted the attention (and the finances) of Japan's leading business houses. Aid projects financed by loans were very much business ventures for those companies involved.

Government loans were given at better-than-commercial terms and were intended to assist economic growth and welfare in the developing countries by supplementing the domestic resources of LDCs. The purpose of a loan varied according to the current level of economic development, but on the whole it was economic and social infrastructure projects which were financed. Commodity loans were also given. In 1977, for example, official direct loans formed 12 per cent of total official and private cooperation and 46.5 per cent of ODA. The emphasis on loans policy was heavily on project assistance, 75.2 per cent of total loans in fiscal 1976 being for projects. 19.5 per cent was devoted to commodity loans and 7.3 per cent to debt relief.[34] In June 1977 agreement was reached between the members of the interministry loans committee on a set of general principles to govern loan commitments: while project loans would remain the basis, commodity loans would be encouraged for balance of payments support and other purposes; capital equipment loans would be considered; loans to relatively developed countries would continue; renewed efforts would be made to select good projects; new forms of lending ('fixed frame' loans, local cost financing, joint financing with international institutions) would be assessed on their merits.[35] This was to be consistent with the aim of promoting the trend to some 'social infrastructure' development through loans, provided a proper balance was retained with economic infrastructure projects, and that projects funded were 'sound'.

These concrete initiatives were welcomed by critics of Japanese aid, but they still proved vague enough to allow for their cautious application. The implementation stage was all-important: unlike *gratis* assistance, a four-ministry committee made decisions on official loans, which the OECF (or the Eximbank in non-ODA assistance) carried out. The standard pattern of request-assessment-decision-implementation was followed, the participants were different, the problems more far-reaching, the sums larger and, therefore, the stakes for both donor and recipient higher. No one ministry was pre-eminent and no one policy position predominated. Equal partnership came closest to being realised although policy was, as even the 1977 guidelines accepted, very much the sum of project-by-project decisions.

In loans, as in grants and technical assistance, requests for aid were, in principle, the first stage of the bureaucratic round. Chapters 6 and 7 describe, however, how decisions could be pre-empted and options limited, how requests and project proposals often came towards the end of tangled and informal lobbying, the product of ongoing bilateral relations. Procedures were inviolable none the less and rules had at least to be seen to be satisfied, so a government-to-government request was deemed necessary. This applied even to political level aid discussions, for requests regularly emerged from ministerial and other contacts between Japan and the developing nations. These approaches (such as those made when Miki Takeo visited the Middle East as special envoy at the time of the oil crisis, or those which came up when Prime Minister Bandaranaike of Sri Lanka visited Japan in November 1976) were anticipated by the Japanese, but the government required that proposals should still be sent through official channels before the project could be properly assessed and approved.[36]

Requests (except those involving a donors' consortium[37]) were presented in the first instance to the Japanese mission in the potential recipient country. Officers there studied the proposal, although as part of their duties they were expected to have made prior representations to the recipient government to outline Japanese policy. It was seen as important to make it clear to LDC governments that the following ODA loan proposals were unacceptable: those related to military activities and provisioning, or to housing, education, research, ships, aircraft or consumer goods. Mission diplomats were careful to give no undertakings on behalf of the Japanese Government and their contribution was restricted to forward intelligence and advisory functions. Although this input was essential, decisions were the prerogative of the ministries at home.[38]

Proposals were then forwarded to the MFA in Tokyo, where the government had firstly to decide whether or not to accept the proposal, a decision based on MFA policy and on the result of interministry discussions. The MFA could reject the request outright after informal contact with other ministries, but if this occurred it was usually for reasons relating to general acceptability and such refusals were clear cut.

The MFA drew up its own policy on the proposal in question prior to the four-ministry conference which the Economic Cooperation Bureau's First Economic Cooperation Division convened and chaired. This division was the nub of loans policy-making, although its institutional standing was not always matched by its influence. Development of MFA policy lay formally in the hands of the division, but was restricted by the authority and the attitudes of regional bureaus and, to a lesser extent, the Policy Division of the Economic Cooperation Bureau. Likewise, the Treaties Bureau was always consulted when loan agreements were being considered. The political and diplomatic implications of bilateral loans were always closely watched by officials. The MFA's 1978 aid report lists political considerations as one of the four sets of factors normally assessed by donors when distributing aid. Under this heading were included geographical proximity, social and cultural affinity and political familiarity. Other sets were the level of poverty of the LDC in question, its ability to use aid effectively and specific economic data (trade potential, natural resources and so on).[39] Regional bureau interests were therefore inextricably implicated, although more so in the case of the major loan recipients, for the bureaus' resources did not allow them to monitor closely the aid relationships with all the countries for which they had responsibility. Officials of these bureaus were nevertheless adamant that theirs was the influential voice on loan aid.

Doubts about the advisability of refusing a request were resolved in initial interministry discussions. These formally took place only between the four ministries represented on the committee, and the exclusiveness of this group was staunchly defended by the incumbents. Although 47.9 per cent of total accumulated loan aid up to September

1977 was spent on agriculture, transport, communications and water supply projects, the MAFF and the Ministries of Construction, Transport, Posts and Telecommunications and Health and Welfare were excluded from the committee. This no doubt made debate more manageable and (given the notorious inability of even the four participating ministries to agree) less drawn out, but the increasing share of these sectors in the loan programme led to demands from these other ministries for channels to give advice more open than a place on an assessment team or informal lobbying of sympathetic MFA or OECF/ EPA officials. In any case, the MFA's First Economic Cooperation Division circulated to the MOF (Overseas Public Investment Division) and the EPA (First Economic Cooperation Division) details of the request and, if it had already undergone feasibility studies, the results of these. Requests needing feasibility studies were rerouted to technical assistance divisions to be incorporated in their priority lists, if considered appropriate. While Chapter 6 will detail the important place of surveys in the aid process, it is necessary to point out here that a decision to go ahead with a project loan was contingent on completion of a feasibility study and on LDC-government approval of the feasibility report. Only on highly political requests could a loan or loans be committed before a feasibility study was made: aid promised to the Middle East in 1974 was an example, as was 'fixed framework' financing where aid was pledged for a fixed period or purpose and projects decided later. Aid to Indonesia through the Inter-Governmental Group on Indonesia (IGGI) enjoyed this freedom, and so did commitments to Papua New Guinea in December 1977, but the Japanese Government was reluctant to use it more widely. The much-vaunted ASEAN projects were not given this assurance when Mr Fukuda pledged Japanese assistance in August 1977.

Interministry committee meetings were held initially at deputy division director level and, if necessary, at division director level. About one-third of cases were approved by the division directors after desk officer meetings had been held. Most others required a conference of division directors and sanction by more senior officers. Only a very few proposals, for large and important projects or for aid to recipient countries with which Japan had a special or sensitive relationship (Indonesia, South Korea and some Middle Eastern nations), demanded meetings of bureau directors. Division directors, therefore, were the officers with the strongest and most direct power over loan decisions.

At the first meeting the MFA representatives offered their position paper, outlining the request and its background and setting out the suggested amounts, terms and conditions of the loan. They presented arguments on the technical aspects of the proposed project and on the economic and political implications of the request. While early responses from other ministries, especially the MOF, were guarded, each had an opportunity to put its point of view at these early meetings. There were few surprises, however, for ministries had had ample time to prepare their cases, particularly if feasibility studies were already completed. Prior informal talks between responsible desk officers ensured predictability in negotiations but agreement was never guaranteed.

All ministries weighed requests in terms of their own priorities but there were (apart from the more general considerations mentioned above) three common to all:

1. Suitability of the requesting country as a recipient, which involved analysis of its level of economic development and of its present and future political and trade relations with Japan. Judgements were made on its development potential (based on *per capita* income), its regional influence, the extent of diplomatic (treaties, degree of cooperation in international bodies) and emigration ties with Japan, its importance as a source of raw materials and as a market and on the state of its trade balance, international payments and debt burden.
2. Appropriateness of the particular project, which required considering the extent to which the request fitted the loan conditions of the various agencies, its priority in LDC development plans, the contribution which Japanese assistance could make to the project and to general relations with the country, and of the prospects for the project's being properly completed.
3. Project details: feasibility report, project content, cost, plans for materials' procurement etc. were at issue and were not determined until after the feasibility study was finished.

The MOF point of view was a telling factor in policy towards a request. This was because both the MOF International Finance Bureau and Budget Bureau prepared the MOF position on the request; the often favourable attitude of the former could be restrained by the domineering fiscal rigidities of the latter. The International Finance Bureau's Overseas Public Investment Division was the official channel to the Budget Bureau, which studied carefully the financial conditions of the loan, such as quantity, terms, repayment period, Japan's balance of payments and currency reserves and the future impact of the loan on OECF budgets. It was possible for the MFA to make informal representations directly to the Budget Bureau, although the International

Finance Bureau did not look kindly on being bypassed. As it was (on aid) ideologically placed between the opposite poles of the MFA and the Budget Bureau, its often helpful middle-man role was easily undermined by unofficial negotiations. The 'constructive competition' between ministries was easily jeopardised.

The OECF was also involved at a formal and an informal level. Its officials attended initial desk officer meetings as observers, but were not present at higher discussions. Their opinions about the technicalities of projects and requests were frank and forthcoming, but did not extend to advice about types of projects or recipients. These were policy questions outside an implementing agency's brief but, informally, officers of the fund and the ministries mixed easily at a professional and social level, especially because many senior fund personnel were originally from the ministries. Continual turnover in ministry aid division staffs also gave OECF officers the edge in know-how, experience and judgement on projects or on details of loans policy, a point which will be taken up in Chapter 8.

Once agreement between ministries was reached, the MFA began talks with the recipient government, leading up to an official exchange of notes, which signified a Japanese Government commitment. The Japanese note was drafted in the First Economic Cooperation Division, checked in the Treaties Bureau (which, as Fukui notes, 'exerts substantial unifying influence' on MFA policy-making[40]) and shown to other ministries. Negotiations were in most cases undertaken in the recipient country by the Japanese mission. Cabinet approval of the exchange of notes followed, after which a loan agreement was concluded between the recipient government and the OECF. Project preparation and payment of the loan were carried out in accordance with provisions of the agreement.[41]

Because four ministries contributed to decisions, bureaucratic initiative in loans policy was restricted. Responsibility rested with a committee, not with any one division. Loans policy was hard to quantify, built up on a case-by-case basis. There was no planning for loans policy as occurred, in a minor way, for grants or for technical assistance, except in that ODA disbursement targets required certain levels of commitment and implementation. Nor was there staff for country programming as a first step to a more orderly approach to geographical and sectoral distribution of loans. The OECF had the manpower to assess sectoral aid, but no staff to tackle the country studies it saw as necessary. Guidelines were broad, even vague, and loans performance was hampered by delays at project sites, cost overrun and by the capacity of the OECF to implement efficiently. Poor disbursement performance in 1976, for example, was put down to low commitments (due to fiscal difficulties), delayed projects, a fall in commodity disbursement, slow decision-making processes and cost overrun, where inflation caused a slowdown of or even halt to certain projects.[42] Fiscal 1978, however, saw a disbursement rate of 88 per cent compared with the 66 per cent in 1976, an improvement due largely to an increase in loan agreements and government goals for doubling ODA.[43] However, the officials, as later chapters will show, were in a sense unable to control aid flows as they might have wished; they only came in towards the end of a long series of steps to begin a loan project. Their decisions were swayed by their perceptions of the proposed project, and at this important point of control there was little consensus between ministries, for each had its own interests to guard and objectives to pursue.

Aid Policy Coordination

Aid policy was piecemeal. Procedures plainly worked against the development of consistent and mutually reinforcing policy for the different kinds of aid. Each had its own group of officials and, more significantly, its own schedule and momentum, but the directions of loans, grants and technical aid were not necessarily in harmony, for there was no means of consciously producing policy acceptable to all ministries or all officials, except at the lowest common denominator of agreement, which was too imprecise for working rules. Policy revolved around jurisdiction and coordination, both within and between ministries, although each separate aid process was subject to those problems to varying degrees.

In areas where policy-makers were distinct from policy-implementers — loans, technical assistance and to some extent grants — arguments about responsibility and the best policy course were loudest. This was because, in aid policy, implementation affected the size, quality and direction of future aid flows. Aid policy did not end with payment; in one sense, it was only beginning at that point. Where implementation was removed from the direct control of ministries, problems of territory were heightened as officials competed to achieve as great an influence as possible over the original decisions and, indeed, the flow of aid requests. Technical aid policy was rife with issues of competence, and loans policy-making was structured on the premiss that differences between ministries could not be resolved. Arguments within ministeries also affected policy outcomes and, although usually more manageable than those between ministries, were a permanent feature of certain procedures.

Coordination was not placed above other ministry priorities. Emphasis on procedures concentrated attention on the separate stages of the policy process and on participants, but not on goals. Officials relied heavily on budgeting to draw elements of aid policy together, although Chapter 5 will demonstrate that this was not effective. Even aid subject to attempts at planning, such as technical aid, could not be extracted from the short-sighted constraints of budgeting. Loans policy was largely isolated from decisions made about other forms of aid.

The nature of aid policy as one encroaching on significant areas of government — tariffs, trade, industrial structure, for example — and, in Japan, a policy with no fixed bureaucratic 'home' and no steady political support, meant that fundamental changes in aid policy depended on the extent to which coordination could be effected (a) between aid and other national policies; (b) between the kinds of aid to a particular country; and (c) between different parts of the entire aid programme. This coordination could not come easily, nor quickly, for in all of these categories there was a prior need for a detailed government statement of objectives in aid policy, some set of principles to underpin ministry programmes. This did not exist in 1978.

Initially, an attempt had to be made to rationalise the relations between aid and other national policies, not to supplant them but to find an appropriate marriage of the policy interests of the main ministries. Accommodation *between* ministries, however, was premised on consistency *within* ministries and this proved difficult in many areas. Coordination at the national level required some compromise between the positions of the MFA, the MOF and MITI, but here the MFA was at a disadvantage, for it had no domestic power base, was unskilled at domestic bureaucratic infighting and had few weapons to use in budget negotiations. The absence of a professional aid service further inhibited the development of a coordinated aid policy.

If high-level political support were not forthcoming, did a 'softer' form of coordination take place lower down in the bureaucracy, to assist aid policies towards individual countries? Coordination, in the words of Heclo and Wildavsky, was 'knowing what other people . . . are saying and doing . . . fostered by a functional redundancy, as it were, of overlapping, criss-crossing and repetitive channels of communication'.[44] Precise information about countries, projects and possibilities was vital if aid requests were to be assessed quickly and objectively, but most officials admitted that, while there was much constructive informal contact between aid officials throughout government (especially because of the frequency of the informal meeting in daily work patterns), there was insufficient exchange of substantive information about each other's ideas and tasks. Survey reports, for example, were not circulated widely enough to prevent feasibility studies being duplicated and policy decisions about requests being delayed. The extent to which policy-making for any form of aid was monopolised by one ministry determined the access other ministries had to information. Furthermore, distinctions between the aid needs of small and large recipients were blurred. The inflexibility of procedures (and the absence of aid specialists) hindered officials in creating programmes suitable for minor recipients and in establishing new directions for policy, by forcing the aid programme as a whole to follow standards laid down for major recipients.

Coordination between different types of aid was based on (a) general policy priorities, (b) the patterns of aid to large recipients, and (c) the extent to which bureaucratic controls (budgets, interministry committees etc.) offset the rigidity imposed by the spread of functions. Such measures had not worked well in the past and budgeting was often harmful in prolonging decisions between ministries when aid budgets were involved, and building into the system — by allowing incrementally increasing aid budgets — barriers to the matching of ministry functions (as the next chapter will demonstrate). Budget requests could only be harmonised if the administration were centralised to some degree. Calls for the creation of a central aid agency became fewer, however, and while interministerial committees were used, they tended to ensure that ministry views were merely represented in the formulation of loans policy; they did not lead to diverse ministry programmes being assimilated.

Notes

1. George Cunninham lists the major aid decisions facing donor governments as: distribution of funds between recipients; choice of project or programme aid; active or reactive role on aid utilisation; extent of country programming; selection of projects; forms of committing funds; terms of aid; total volume. See his *The Management of Aid Agencies: Donor structures and procedures for the administration of aid to developing countries* (London, Croom Helm, 1974), pp. 1-30.

2. Gaimushō keizai kyōryokukyokuchō-hen, *Keizai kyōryoku no genkyō to tembō: namboku mondai to kaihatsu enjo* (Director- General, Economic Cooperation Bureau, Ministry of Foreign Affairs (ed.), *Economic Cooperation: present situation and prospects: the North-South problem and development assistance*) (Tokyo, Kokusai kyōryoku suishin kyōkai, 1978, hereafter *Namboku mondai*), pp. 400-1. For details of Japan's grant aid, see pp. 387-412, and Miyake Kazusuke, 'Kokusai kyōryoku no genjō (6)' ('The present situation in international co-operation (6)'), *Kokusai tsūshin ni kansuru shomondai*, vol. 23, no. 8 (November 1976), pp. 23-8, and Tanimura Yorio, 'Nikokukan mushō shikin kyōryoku' ('Bilateral capital grant cooperation'), *Keizai to gaikō*, no. 647 (April 1976), pp. 58-61.

3. The Japanese Government appreciated the need to move towards local cost financing of capital grant projects, in line with DAC and UNCTAD IV resolutions and also recognised that greater untying of grants was desirable. LDC untying of capital grants was possible, although suppliers could only be Japanese or Japanese-controlled firms. A commitment to untie grant aid to the poorest countries as a form of debt relief was made in March 1979 (see *Japan Times Weekly*, 17 March 1979).

4. *Namboku mondai*, pp. 769-70.

5. See Kokusai kyōryoku jigyōdan mushōshikin kyōryokuka, *Mushō shikin kyōryoku* (Japan International Cooperation Agency, Capital Grant Cooperation Division, *Capital grant cooperation*), n.d., p. 3.

6. Miyake, 'Kokusai kyōryoku no genjō (6)', p. 38.

7. One such example cited by MFA officials was the refusal of a request from a small African country for a Japanese make of electron microscope.

8. As with the reparations-like grant made to Mongolia of 5,000 million yen in 1977, which appeared to have been the initiative of officials in the China Division of the Asian Affairs Bureau and the Treaties Bureau. For details, see *Asahi shimbun*, 21 November 1976.

9. Of grants committed from 1969 up to the end of fiscal 1974, South Vietnam received 13,158.17 million yen or 50.1 per cent of the total, well ahead of Bangladesh, which received 5,582.6 million yen or 21.3 per cent. In the years 1975-September 1977, the Socialist Republic of Vietnam received Japanese grants totalling 13,500 million yen or 29.1 per cent of the total. See *Namboku mondai*, pp. 392-5.

10. In 1976, for example, 28 grants for specific purposes were committed. Up to 1975 the numbers were smaller, between 2 and 12.

11. General capital grants, excluding food aid, were budgeted in fiscal 1975 the amount of 15,763 million yen, fiscal 1976 16,000 million yen, fiscal 1977 18,000 million yen, but a 217 per cent increase to 39,000 million yen in fiscal 1978.

12. *Namboku mondai*, pp. 392-5.

13. Although there were provisions for the carry-over of grant funds, the disbursement of budgeted amounts was low. Audit figures showed that percentage disbursement in, for example, 1969 was 27 per cent, 1970 43 per cent, 1971 75 per cent, 1972 75 per cent, 1973 23 per cent (fiscal years). This was in comparison to disbursement of total MFA aid budgets in the same years of 86 per cent, 85 per cent, 90 per cent, 88 per cent and 61 per cent. See Ōkura zaisei chōsakai, *Shōwa 47-51 nenpan: kessan to kaikei kensa (1972-76 editions: settlement and audit of accounts)* (Tokyo, 1971-5), covering the 1969-73 budgets. *Namboku mondai*, p. 696, has figures for 1974-6: 43, 24 and 59 per cent respectively.

14. *Namboku mondai*, p. 505.

15. An excellent insight into the Finance Ministry is provided in John

Creighton Campbell, *Contemporary Japanese Budget Politics* (Berkeley, University of California Press, 1977).

16. *Nihon keizai shimbun* (hereafter *Nikkei*), 23 November 1976.

17. Aid consortia were dealt with by the First Economic Cooperation Division of the Bureau, since they were part of bilateral aid relations.

18. Huang Po-wen Jr, *The Asian Development Bank: Diplomacy and Development in Asia* (New York, Vantage Press, 1975), pp. 17-19.

19. F.C. Langdon, *Japan's Foreign Policy* (Vancouver, University of British Columbia Press, 1973), p. 172.

20. John White, *Regional Development Banks* (London, ODI, 1970).

21. Michael Haas, 'The Asian Development Bank', *International Organization*, vol. 28, no. 2 (Spring 1974), p. 292.

22. Interview, 8 June 1976.

23. Interview, former EPA official, 8 July 1976.

24. See his 'DAC tai-nichi nenji shinsa ni shusseki shite' ('On attending the annual DAC examination of Japan'), *Keizai to gaikō*, no. 628 (September 1974), pp. 50-5 at p. 55.

25. It was claimed by an examining country official that on one occasion when Japan was to be examined by the DAC, MFA officers from the Japanese OECD mission circulated lists of questions which they wanted the examining and other country representatives to ask, primarily to highlight the problems which the MFA wished pointed out to the MOF officials present.

26. Interview, 7 October 1976.

27. There is no detailed and extensive analysis of Japanese technical assistance policy available in English. The MFA report, *Namboku mondai*, pp. 413-39, gives a concise account in Japanese. Successive JICA annual reports have more detail.

28. DAC documents on 1978 aid disbursements.

29. Between 1954 and 1976, 24.6 per cent of trainees accepted were in the agricultural sector, 12.3 per cent in mining and industry, 10.2 per cent in communications, 8.8 per cent in transport and 6.2 per cent in the construction sector. For specialists and survey teams sent overseas, the figures for the same period were 20.3 per cent, 13.2 per cent, 7.1 per cent, 9.5 per cent and 16.7 per cent for specialists and 6 per cent, 19 per cent, 6.8 per cent, 16.7 per cent and 29.8 per cent for development surveys. See Gaimushō kanshū, *Me de miru gijutsu kyōryoku* (Ministry of Foreign Affairs (ed.), *Technical cooperation illustrated*) (Tokyo, Kokusai kyōryoku suishin kyōkai, 1977), pp. 46-52.

30. This question is taken up again in Chapter 8. Judith Tendler discusses project generation by development assistance institutions in her *Inside Foreign Aid* (London, The Johns Hopkins University Press, 1975), especially pp. 103-4.

31. Cunningham, *The Management of Aid Agencies*, Ch. 1. There was no legal requirement for requests, but it was a practice long used by the Japanese Government, and many officials would have preferred a more overtly active approach.

32. Iida Tsuneo, in his *Enjo suru kuni sareru kuni (Aiding and aided nations)* (Tokyo, Nihon keizai shimbunsha, 1974), pp. 18-19, gives an excellent example. It seems that in order to have Iida sent to Indonesia to assist in economic planning, an Indonesian Government request on a specified request form was necessary, despite unofficial agreements on his going. Embassy officials drew up the form and waited outside the office of the appropriate Indonesian minister to obtain his signature when he emerged. Even then, Iida was not expected when he eventually turned up.

33. *Namboku mondai*, pp. 418ff.

34. *Namboku mondai*, p. 450. I shall restrict my discussion here to OECF loans which, since July 1975, have been the only loans given with a grant element

of 25 per cent or above. Eximbank finance, by agreement between the two agencies, was restricted to less concessional loans and finance to Japanese companies. See *Nikkei*, 20 June 1975, reporting *Yūgin-kikin no gyōmu bunya chōsei ni kakaru oboegaki ni kansuru ryōkai jikō* (Terms of understanding reached in the memorandum on the coordination of the responsibilities of the Eximbank and OECF), 20 June 1975.

35. *Namboku mondai*, pp. 767-8.

36. One case which did not go through proper channels was the oil loan agreed to by the Prime Minister, Satō Eisaku, in 1972 in exchange for Indonesian guarantees of oil for ten years. There were numerous rumours about the propriety of this loan, suggesting personal deals between President Suharto of Indonesia and Satō. See *Nikkei*, 14 May 1972; Kitazawa Yōko, 'Nihon-Indonejia no seiji keizai kankei' ('Economic and political relations between Japan and Indonesia'), Part 1, in *Keizai hyōron* (*Economic Review*) (September 1976), pp. 84-99 at pp. 96-7; and Ogawa Kunihiko, *Kuroi keizai kyōryoku: kono ajia no genjitsu o miyo* (*'Black' economic cooperation: the Asian situation*) (Tokyo, Shakai shimpō, 1974), Ch. 6.

37. Japan was a member of a number of aid consortia, such as those for aid to India, Pakistan, Bangladesh and Indonesia.

38. Most officials saw the mission role as a vital one for intelligence and communication, given its knowledge of LDC conditions. There is further discussion of this in Ch. 8.

39. *Namboku mondai*, p. 757.

40. Fukui Haruhiro, 'Policy-Making in the Japanese Foreign Ministry' in Robert A. Scalapino (ed.), *The Foreign Policy of Modern Japan* (Berkeley, University of California Press, 1977), p. 18.

41. On normal project aid the exchange of notes was followed by a legal questionnaire from the OECF to the recipient government, before further study of the project (and on-site surveys if the OECF deemed them necessary) could be carried out. Loan agreement negotiations were entered into, after which the loan agreement was concluded. Following this, there were several kinds of payment procedure which could be used, depending on the country in question. The most common was the 'commitment' or 'letter of credit (L/C) switch' method, where letters of credit were exchanged. Some countries did not allow the use of letters of credit, in which case the 'direct payment' procedure was adopted. The 'reimbursement' method was also possible.

42. See Gaimushō keizai kyōryokukyoku keizai kyōryoku dai-ikka, *1976-nen no seifu chokusetsu shakkan kyōyo jōkyō (kōkan kōbun teiketsu bēsu)* (Ministry of Foreign Affairs, Economic Cooperation Bureau, First Economic Cooperation Division, *Provision of direct government loans in 1976, exchange of notes basis*), 21 December 1976.

43. *Japan Economic Journal*, 10 April 1979.

44. Hugh Heclo and Aaron Wildavsky, *The Private Government of Public Money: Community and Policy Inside British Politics* (Berkeley, University of California Press, 1974), p. 69.

5 BUDGETING FOR FOREIGN AID

Annual budgeting was, to most aid officials, an inescapable fact of life, which dominated policy discussion over the whole fiscal year from April to March. Budgeting was a feature of aid policy-making in most donor countries. All DAC members set aside yearly a portion of their central government's budget for development assistance, the share of these allocations within the total budget ranging from less than 1 per cent to over 3 per cent. Most Western donors provided nearly all of their aid in this way, but some lessened their reliance on budgetary funds by using other sources. Likewise, there were procedures which affected the flexibility of appropriations, although, commonly, specified disbursements were voted yearly. In many countries, depending on budgetary tradition and the significance of aid in government priorities, there existed provisions for the carry-over of unspent funds, for forward budgeting in the form of advance commitments which could be charged to future budgets, or for indicative medium-term plans.[1]

'Expenditure is policy; policy is expenditure', Heclo and Wildavsky tell us.[2] The power of the purse over policy was entrenched in Japan and, as Campbell argued in his study of Japanese budgeting, 'more decisions were made as part of the budgetary process in Japan than elsewhere'.[3] In other words, more policy decisions were left to 'budgeters'. This was encouraged, Campbell held, by systemic factors, including political stability and the tendency for the rules of budgetary compilation to substitute for the rules of decision-making. Policy disputes were reduced to technical problems and incrementalism prevailed. Several factors — the uneasy relationship which existed between aid and the national government in Japan (as it often did in other donor countries), the active participation of the MOF in administering the aid programme and the restraint imposed by annual budgeting on all sections of the Japanese aid programme — would suggest that aid was no exception. While many aspects of foreign aid in Japan were intimately related to expenditure and budget compilation, and while the MOF was the *bête noire* to many an incipient programme, aid policy-making was not entirely dependent on MOF budget officials. Budgeting affected aid policy but did not wholly determine it.

150

Aid Flows and the Budget

Over half of Japanese flows of funds to the developing countries came from budgetary sources. ODA and Other Official Flows comprised 55 per cent of total flows in 1977, a figure which fluctuated over the decade of growth of Japanese assistance as private investment rose and fell. All officials flows came directly or indirectly from budget funds, and the authorisation of aid expenditure took place when the Diet passed the government's budget bill in March each year for the financial year which began on 1 April.[4] The voting of funds represented official legislative control of both the volume of expenditure and its timing, for any carry-over of moneys needed Diet sanction. The legislative power, however, did not extend to all government aid and, in practice, the passiveness of the Diet in the drawn out and often tedious budgeting process meant that *de facto* authorisation occurred while the budget was being prepared some months before. The real power in determining the aid budget lay with the officials.

Aid was voted through different budget sections (see Figure 5.1) and was not only derived from tax revenues through the General Account. Bilateral and multilateral grants and government capital subscriptions to the Overseas Economic Cooperation Fund (OECF) were all defrayed from the General Account (*ippan kaikei*). They were listed under the 'Economic Cooperation' subsection of the General Account in the MOF's compilation. The Export-Import Bank (Eximbank) received capital from the Industrial Investment Special Account. Both the OECF and the bank were able to borrow from the Trust Fund up to limits specified in the budget each year,[5] and in 1979 approval was given for the OECF to issue its own bonds in order to boost its capital and assist in achieving the aid-doubling target. While the single-year budgeting principle (budgets had to be spent within the one fiscal year) was a time-honoured tradition of the MOF, some aid categories were provided for by being allocated to special accounts for disbursement over several years. Reparations were transferred from the 'Economic Cooperation' subsection to the Special Account for Reparations and other Special Foreign Obligations, while bonds for subscriptions to multilateral financial institutions were managed through the National Debt Consolidation Fund Special Account. Both these accounts allowed multiyear disbursement.[6] There were also provisions for the carry-over of some unspent funds which, in the case of capital grants, was often a considerable proportion of the year's allocation.[7]

The Industrial Investment Special Account and the Trust Fund were

Figure 5.1: Sources of Budget Funds for Official Aid

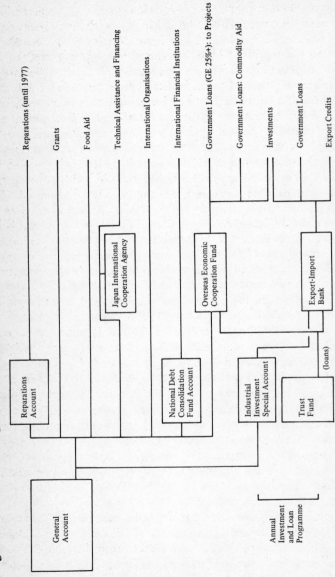

Source: Gaimushō keizai kyōryokukyokuchō-hen, *Keizai kyōryoku no genkyō to tembō: namboku mondai to kaihatsu enjo* (Tokyo, Kokusai kyōryoku suishin kyōkai, 1978), p. 666.

part of the annual Fiscal Investment and Loan Programme (FILP) which was compiled by the Financial Bureau of the MOF in consultation with the Budget Bureau. The FILP, although not based on taxation revenuè and lying outside the immediate scope of the government's 'policy' expenditure, provided 'a significant portion of Japanese investment in economic development and social overhead'.[8] It was, to the same extent, important to the overseas loan programme of the government. Because it was used to support the government's main statutory corporations, and derived most of its resources from postal savings and insurance contributions, the FILP was integral to Japanese fiscal and monetary policy. It was also a useful secondary tool of careful budgetmakers, as they strove to restrict the demands on the tax-based General Account and divert pressures into increased FILP expenditure. As Table 5.1 shows, the share of the official economic cooperation budget which was derived from the Investment and Loan Programme and from the General Account was evenly balanced over the period 1965-75 at approximately 3:1, since increased capital for the OECF from the Trust Fund was offset by a fall in outlays to the Eximbank from the Industrial Investment Special Account. This reflected an enlarged responsibility for the OECF in government loans and, at the same time, an increased percentage of total aid given as bilateral loans.

By 1977 this ratio had dropped back to 4:1, but it rose again to about 3:1 in 1978-9 after large increases in the grant aid budget in both years, plus additional allocations in the 1978 supplementary budget. The budgetary requirements for doubling ODA (originally in five years and later in three years) gave a great boost to OECF subventions from the FILP as well as overall General Account increases for aid. Of course, the ODA budget, a sub-category of the total economic cooperation budget, depended rather more on the General Account, which provided in the region of 50-60 per cent, the rest coming from Trust Fund loans to the OECF and Eximbank and the OECF's own accumulated capital resources. The aid-doubling programme will rely to a great extent on the General Account, although the OECF had its Trust Fund allocations doubled in the 1979 budget as a step towards achieving the government targets.

By conscientiously maintaining a 3:1 balance in the total economic cooperation budget over the years, the MOF, in particular the Budget Bureau, ensured that the drain on the General Account budget caused by economic cooperation did not increase in proportion to Japan's expanding aid effort. The use of loans from the Trust Fund rose in gross terms and as a percentage of the total economic cooperation

Table 5.1: Economic Cooperation Budget, According to DAC Categories, 1965-77 (Fiscal Year, Yen Million)

	1965	1966	1967	1968	1969	1970	1971	1972	1973	1974	1975	1976	1977
1. Bilateral Grants:													
Grants	26177.0	36216.0	39881.9	44460.7	44455.0	43776.9	44935.5	47476.2	56261.1	68847.4	70832.0	64479.0	74496.0
Technical	7264.1	15063.5	16444.1	17312.0	19654.2	18539.6	17356.0	18625.5	23143.8	29059.8	24947.1	19480.0	26150.0
Reparations	2713.0	3752.6	5437.8	6588.7	7742.9	9288.0	11475.4	15015.0	19086.1	25656.6	31638.0	36683.0	43325.0
Food Aid	16200.0	17399.9	18000.0	18000.0	11908.8	10800.0	10800.0	9240.0	9240.0	9240.0	9240.0	3080.0	—
	—	—	—	2574.3	5148.9	5148.9	5148.9	4405.4	4651.6	4713.2	4713.2	4713.0	5021.0
2. Multilateral:													
Grants	1630.4	1857.6	2092.3	2577.2	2629.8	3129.0	3662.4	4408.4	7352.1	9902.5	21524.0	22497.0	21103.0
Subscriptions	8787.6	12150.0	12150.0	14400.0	30427.2	35352.0	32400.0	35356.0	40296.0	69448.2	61784.0	83480.0	132887.0
3. Bilateral Loans:													
OECF	1000.0	7500.0	9000.0	6000.0	22400.0	29000.0	33000.0	42000.0	56000.0	65000.0	65000.0	75500.0	76000.0
Eximbank	29000.0	37000.0	48000.0	48000.0	63500.0	76000.0	65000.0	63000.0	63000.0	60000.0	60000.0	67000.0	63000.0
JICA	—	—	—	—	—	—	—	—	—	7030.0	10198.0	13200.0	17200.0
Other	—	—	—	—	—	—	—	—	—	4310.0	7303.0	9670.0	11256.0
4. Total	66595.3	94723.6	111124.4	115437.6	163411.2	187257.9	183197.9	195740.6	226225.2	284786.0	307691.0	335826.0	395942.0
5. Trust Fund:													
OECF	1000.0	7500.0	9000.0	20000.0	27600.0	31000.0	40000.0	61000.0	69500.0	77000.0	94500.0	97800.0	98000.0
Eximbank	91900.0	115000.0	185000.0	215000.0	282000.0	273000.0	314000.0	420000.0	486500.0	556500.0	559500.0	703600.0	822000.0
6. Grand Total	159495.3	217223.6	305124.4	350437.6	473011.2	491257.9	537197.9	676740.6	782225.2	918286.1	961691.0	1137226.0	1315942.0

1. Eximbank finance was not all provided for economic cooperation (almost all Trust Fund total for non-ODA purposes).
2. Borrowings by the OECF and Eximbank from the Trust Fund were limits only. Not all funds are drawn upon.
3. The large increase in technical cooperation funds between 1973 and 1974 was due to the inclusion of an item previously recorded under multilateral contributions.
4. Eximbank funds for bilateral loans were drawn from the Industrial Investment Special Account.
5. 'Other' loans are those from Japan Overseas Development Corporation, Fisheries Promotion Fund etc.

Source: Gaimushō keizai kyōryokukyoku, *Keizai kyōryoku kankei shiryō* (Ministry of Foreign Affairs, Economic Cooperation Bureau, *Materials on economic cooperation*) 1973, 1974 and 1976. 1965-9 figures were converted from $ equivalents given in the 1973 volume, at a rate of $1 = 360 yen. 1976-7 figures from Gaimushō keizai kyōryokukyokuchō-hen, *Keizai kyōryoku no genkyō to tembō*.

budget, by more than General Account subscriptions to the OECF. This diversionary policy was also applied to the Eximbank, the Industrial Investment Special Account capital of which was itself originally transferred from the General Account. Over the period 1965-77, loans from the Trust Fund to both agencies rose nearly tenfold, far higher than increases in the total General Account. In recent years, the Budget Bureau has expressed a desire to remove the 1:1 balance of borrowings and capital provided for in the OECF Law, in order to increase the Trust Fund allocation and to reduce the impact of the OECF on the General Account. This was resisted by the EPA, the OECF and the MFA — who feared the adverse effect this might have on OECF interest rates — and by the MOF's Financial Bureau, which administered the Trust Fund and did not want any extra burdens placed on its resources. A revision of the OECF Law was passed by the Diet in early 1979, however, raising the ratio of borrowings to capital to 3:1.

The low political priority of foreign aid (compared with other budget categories) encouraged the ever-vigilant Budget Bureau to lighten the expanding aid programme's impact on the General Account by shifting the burden of financing bilateral loans onto the Trust Fund. This in itself was politically important, because the annual Investment and Loan Programme (of which the Trust Fund was the main component) was not subject to Diet approval. Table 5.2 shows the shares of the General Account held by main policy categories, and their annual growth. It is evident from the table that 'Trade Promotion and Economic Cooperation' was a very small percentage of the General Account when compared with other categories, although over the period 1965-74 it rose from 0.4 per cent to 1 per cent. Growth was not steady, however, and since 1969, when 'Economic Cooperation and Trade Promotion' reached a high of 1.2 per cent, its share declined regularly to the point where it was only 0.7 per cent in 1977. Indeed, between 1965 and 1969 it grew by 640.3 per cent, but dropped to 73.8 per cent between 1969 and 1974. The declining share of economic cooperation in the General Account contrasted with the increasing share of social security expenditure over the period, a stable proportion for education and a high, but slightly falling, allocation to public works. It was clear, therefore, that foreign aid in that period was losing in the competition for public grants through the General Account. This may change marginally under the government's aid-doubling plan.

Although cushioned somewhat by the FILP, the economic cooperation budget grew after 1965 in a way completely different from the total General Account. Compared to the growth of the General

Table 5.2: Shares of Various Policy Categories in the Budget's General Account, 1965-77 (Fiscal Year, Thousand Million Yen)

	1965	1966	1967	1968	1969	1970	1971	1972	1973	1974	1975	1976	1977
Social Security	516.4	621.7	719.5	815.6	946.9	1137.1	1344.0	1641.5	2114.5	2890.8	3926.9	4807.6	5692.0
% of Total	14.1	14.4	14.5	14.0	14.1	14.3	14.3	14.3	14.8	16.9	18.4	19.8	19.9
% Increase	—	20.4	15.5	13.6	16.1	20.1	17.8	22.1	28.8	36.7	35.8	22.4	18.4
Education & Science	475.7	543.3	624.6	702.4	805.7	925.6	1078.9	1304.4	1570.2	1963.3	2640.1	3029.2	3429.7
% of Total	13.0	12.6	12.6	12.1	12.0	11.6	11.5	11.4	11.0	11.5	12.4	12.5	12.0
% Increase	—	14.2	15.0	12.5	14.7	14.9	16.5	20.9	20.4	25.0	34.5	14.7	13.2
Defence	301.4	340.7	380.7	422.1	438.8	569.5	670.9	800.2	935.5	1093.0	1327.3	1512.4	1690.6
% of Total	8.2	7.9	7.7	7.3	6.5	7.2	7.1	7.0	6.5	6.4	6.2	6.2	5.9
% Increase	—	13.0	11.8	8.2	14.6	17.7	17.8	19.7	16.9	16.8	21.4	13.9	11.8
Public Works	692.6	880.4	1000.5	1070.0	1206.4	1409.8	1665.6	2148.5	2840.8	2840.7	2909.5	3527.2	4281.0
% of Total	18.9	20.4	20.2	18.4	17.9	17.7	17.7	18.7	19.9	16.6	13.7	14.5	15.0
% Increase	—	27.1	13.6	6.9	12.8	17.3	18.1	29.0	32.2	0.0	2.4	21.2	21.4
Trade Promotion and Economic Cooperation*	12.9	28.1	36.5	48.1	95.5	91.9	101.1	115.2	—	—	—	—	—
% of Total	0.4	0.7	0.7	0.8	1.2	1.1	1.1	1.0	—	—	—	—	—
% Increase	—	118.6	29.4	31.8	98.5	-3.7	10.0	13.9	11.7	—	—	—	—
Economic Cooperation	5.4	17.7	21.4	22.5	72.8	80.2	89.0	102.4	128.8	166.0	176.7	183.1	210.9
% of Total	0.2	0.4	0.4	0.4	1.1	1.0	0.9	0.9	0.9	1.0	0.8	0.8	0.7
% Increase	—	224.3	21.1	5.2	223.5	10.0	11.1	15.0	25.8	28.9	6.5	3.6	7.1
Total	3658.0	4314.3	4950.9	5818.6	6739.6	7949.6	9414.3	11467.7	14282.1	17099.4	21288.8	24296.0	28511.2
% Increase	—	17.9	14.8	17.5	15.8	18.0	18.4	21.8	24.6	19.7	24.5	14.1	17.4
GNP	32650.4	38399.5	45322.1	53368.0	62997.2	73237.2	81446.4	95564.4	115263.1	135920.0	149501.0	167110.0	185425.0
General Account/GNP (%)	11.2	11.2	10.9	10.9	10.7	10.9	11.6	12.0	12.4	12.6	14.2	14.5	15.4

* In 1973 'Trade Promotion and Economic Cooperation' was restructured as 'Economic Cooperation'. Figures in the latter category in the period 1965-72 represent the economic cooperation component of 'Trade Promotion and Economic Cooperation'.

Source: *Zaisei chōsakai, Kuni no yosan (The nation's budget)*, various issues.

Account and its other categories, the economic cooperation vote increased erratically. This suggested that the weak domestic impact of economic cooperation weighed heavily on MOF thinking, and that problems such as the balance of payments and foreign exchange reserves influenced economic cooperation more directly than they did other areas of the budget; these factors were prominent when ministries reviewed particular aid relationships and accordingly came into play when the budget was being compiled. Successive government budget explanations always emphasised them, but the politics of budgeting for aid took officials well beyond the rational assessment of the main economic indicators.[9]

How the economic cooperation budget grew was naturally affected by its size within the total budget. Economic cooperation usually occupied less than 1 per cent of the General Account and increases in the late 1960s did not influence the broader dimensions of the government budget, although MOF officials were ever cautious about the possibility of rapidly expanded aid. Larger FILP allocations for economic cooperation were mainly for the Eximbank's export promotion programme, a form of direct assistance to Japanese industries rather than developing countries. Japan became a large aid donor while herself striving to achieve high economic growth and improved standards of living, goals which demanded and received priority in the budget. Even as these pressures eased towards the end of the 1960s, however, there was no steady increase in total aid budgets and certainly no improvement in aid's policy standing represented by its share of the General Account.

Because of the close links between budgeting and the continuing implementation of aid and because the MOF refused to accept arguments for the need to indicate future growth in the aid budget, annual increases were until recent years tied to growth in some individual programmes. This was the case in the late 1960s, for example, when aid budgets began to expand more quickly. The 1966 budget included a new item for subscription to the Asian Development Bank and enlarged funds for the OECF. In 1967 there were new outlays for technical aid, and in 1969 food aid, capital grants and increased OECF funds more than doubled the economic cooperation budget. In the 1970s, however, the large fluctuations in the economic cooperation budget flattened out and increases were derived mainly, but not entirely, from incremental additions to the majority of budget items and from across-the-board rises in ministry shares. Aid allocations from the General Account were in 1977 divided among 14 ministries and

Table 5.3: Economic Cooperation Budget, General Account: Changes in Ministry Shares, 1965-77 (Fiscal Year, Percentage)

	1965	1966	1967	1968	1969	1970	1971	1972	1973	1974	1975	1976	1977
MITI	13.8	4.8	4.4	4.3	1.5	1.7	3.1	3.4	4.0	4.4	3.7	3.8	4.3
MFA	67.9	32.4	36.7	37.3	13.5	14.6	17.8	20.1	22.7	34.3	39.4	40.8	45.0
MOF	18.4	62.8	58.9	58.4	84.9	83.7	78.4	75.8	71.6	59.1	52.3	50.2	45.7
MAF							–	–	1.0	1.6	2.4	2.7	2.6
Education							0.7	0.7	0.7	0.7	0.9	0.9	0.9
Welfare											1.4	1.6	1.5

A dash indicates a negligible percentage.

The figures on which these percentages are based were not strictly for ODA. For example, the MITI allocation included funds for JETRO after 1970.

Source: As for Table 5.2.

agencies, each of which made its own budget request and disbursed those funds, although some of the MFA and MITI budgets were passed on to JICA, and some MOF requests (such as reparations) were actually managed by the MFA. Table 5.3 shows those ministries which received the largest aid budgets between 1965 and 1977.

The table reveals no obvious pattern in the annual percentage increases for the three main recipient ministries (the MFA, the MOF and MITI), because of the irregular rise and fall of allocations to large programmes. The Asian Development Bank subscription swelled the MOF budget in 1966 as did food aid in 1968, the MFA received a substantial initial amount for capital grants in 1969 and the MITI budget was boosted by assistance for LDC market development in 1971. While growth was uneven, the balance between ministries was maintained, especially after 1970 as shares stabilised and share changes were regularised. With new ministries beginning aid programmes in the seventies and thereby intensifying competition for funds — the MOF did not necessarily subscribe to the 'expanding cake' expectations of the line ministries — shares diverged and the monopoly of the MFA, the MOF and MITI began to weaken. The most apparent change was between the MFA and the MOF, with much larger annual increases in the former's budget, principally because of dwindling reparations payments which had always been attached to the MOF budget. While expenditure by the MFA was consolidated, however, the total aid budget was spread further across other ministries, whose programmes grew steadily, although from a small base. This represented both a strength and a weakness for the MOF: with more participants needing year-round attention (a service which the MOF was not prepared to let slip), its coordinating and organising of the aid effort were bolstered, but its influence on decisions affecting aid commitment was dissipated. Larger macrobudgeting responsibilities led to more complex microbudgeting.

Budget Requests

The power of the Budget Bureau lay in its accepted authority in budgeting, but it did not go unchecked — negotiating by definition implies that unilateral power is diminished. This restrained power, then, was exercised not only in the one-to-one (Budget Bureau to ministry) relationship of the budget request but also in a pattern occurring throughout the year, which reached a peak at budget compilation as the extent of aid disbursement in the fiscal year became apparent. The MOF, including its Budget Bureau, was busy constantly monitoring commitments of loans and grants and the disbursement of grants. This persistence was irksome to many officials in other ministries but was essential if moneys were to be paid. Budget requests for bilateral capital grants were for a fixed amount expected to be committed and then disbursed in the following financial year, whereas requests for multilateral grants and technical assistance (for JICA) included expected disbursements plus an allowance for further commitments. Requests for loan moneys (for the OECF and the Eximbank) were for anticipated disbursements (the MOF having previously assented to government loan commitments formalised in an official exchange of notes) and for small amounts for prospective commitments. In compiling the budget in September to December each year, the scope for the MOF to affect aid commitments as such was limited mainly to grant aid; its ability to influence the committal of loans was exercised chiefly in the ongoing four-ministry loans committee.

Ministry aid budgeting followed the usual hectic budget timetable. The Japanese fiscal year runs from 1 April each year to 31 March of the following calendar year, and ministries began formulating the next year's requests soon after the fiscal year commenced. Over the summer, ministries worked to develop new policies and initiatives and bureau and divisional requests were arranged to form the official ministry application, which was presented to the MOF by the end of August, normally after senior ministry officials had discussed the request with divisions of the Liberal Democratic Party's Policy Affairs Research Council (PARC). This marked the early stage of budgetary lobbying by ministries and the government party.

Ministry explanations of requests to the Budget Bureau took place in September. In turn the Budget Bureau compiled draft allocations for each ministry over October and November, a period of determined activity, late nights and endless cups of tea. In December, the MOF Budget Conference ratified the Draft Budget, which was released about mid-December. This was followed by a week or so of 'appeals (or revival) negotiations' between ministries, the LDP and the MOF − the final, intensely political, step in budget discussions. Cabinet then approved the Draft Budget, which was sent to the Diet as the 'Government Draft'.

Ministries acted well ahead of this timetable to prepare their submissions. Forewarned, to the aggressive aid division, was indeed forearmed. Early talks between ministries and the Budget Bureau took place in May and June, before government policy on the budget requests was decided and before ministries had settled on their own priorities. This allowed both sides to appraise quietly each other's needs and assess the extent of flexibility in the eventual position each would take. Competition for funds, however, did not preclude good manners. Negotiations were friendly, carried on in an atmosphere of earnest intent, the emphasis on both sides being on fairness and direct dealing, in a wider context of accepted (and anticipated) budgeting strategy. The rules were known, understood and obeyed; mutual trust and confidence led to a certain predictability in approach and, usually, response. The Budget Bureau could generally gauge the size of economic cooperation requests even at this preliminary stage, as a result of discussions at two levels.

On the one hand, official talks were held between the bureau and the Accounting Divisions of the respective ministries and, on the other, between the bureau and its client Economic Cooperation Divisions in the course of routine briefings on aid implementation. Both Budget Bureau and supplicant ministries could feel the lie of the budgeting land, and would not hesitate in testing it. The MFA Economic Cooperation Bureau, for example, frequently pressed for greater planning in aid budgeting;[10] the MFA and other ministries had done so, in fact, since as far back as 1961. In May 1976 it put to the Budget Bureau a scheme premised on growth in Japanese ODA to a conservative level of 0.33 per cent of GNP, a target set by the 1976 government Economic Plan, and on substantial improvements in the disbursement rate of Japanese aid committed.[11] Its proposal involved an increase of the share of economic cooperation in the General Account of the budget from 0.82 per cent in 1977 to 0.97 per cent in 1980, a demand which the Budget Bureau indicated it was not prepared to meet. 'There are limits', said one examiner, 'to how far these things can go.'

This early manoeuvring, in which the MFA often tried to secure acceptance in principle of forward targeting for aid budgets, worked two ways, for the same documents could be drawn upon by both parties. One example was a World Bank report cited as being too negative by the MFA in its 1976 paper on budget planning: the report projected Japanese ODA as 0.18 per cent of GNP by 1980, a figure which prompted a DAC response that Japan's 1980 ODA would be 0.36 per cent of GNP. The MOF, however, regarded the World Bank analysis as 'appropriate' and was careful to remind the MFA that the report implied slower-growing aid budgets.

Such arguments — which were common — represented tactical

positioning rather than desperate Budget Bureau attempts to cut back expected requests or MFA moves to undermine the Budget Bureau stance. Both ministries were realistic; they knew, for example, the content of the World Bank and DAC reports in question, and the tenuous assumptions about future GNP growth on which each was based. The more effective budgeting tools were the compiling of the intraministry budget request and the preparation of MOF budget policy for the year. Accounting Divisions could usually sense the ceilings within which the MOF expected them to operate. While the 1977 budget compilation policy was not approved by Cabinet until 30 July 1976, when it imposed a limit on 1977 budget requests of 115 per cent of 1976 allocations, one ministry Accounting Division Director, in early July, already assumed that Budget Bureau attitudes and the prevailing economic climate gave little hope of more than the 115 per cent imposed for the 1976 budget. Of course, it was left to the discretion of this officer, the main budgeting official in ministries, to fit the various components of the ministry application into the request ceiling. The economic cooperation budget request, according to one director, would obviously go above 115 per cent but this could be balanced out within the total ministry figure.

The government's decision in 1977 to double Japan's ODA by 1980 forced on a reluctant MOF a measure of at least indicative budget forecasting for aid (even though it would not accept the need for actual planning) to ensure that the enlarged disbursements to achieve this goal were matched by appropriations. As pointed out in Chapter 1, this would involve doubling US dollar, not yen, values of 1977 disbursement. This lessened the immediate impact on the budget, but it still meant pushing up the ODA budget by between 200 and 300 per cent in three years and improving disbursement rates. Here, of course, lay the crux of the dispute over how aid-doubling should proceed: the MOF refused to admit to the principle of (even rolling) budget plans, which would take aid outside the single-year budgeting framework, and insisted that the emphasis be placed on improving disbursement first. Ministries, on the other hand, had a minor victory when, because of the aid-doubling policy, Cabinet agreed in July 1978 to allow ministry requests for 1979 aid to exceed the limit of 13.8 per cent.

Ministry requests originated in each division and in the implementing agencies but, as was pointed out in the previous chapter, the nature of requests depended on the aid which divisions administered and the effect of the budget on commitment and disbursement. All divisions involved with economic cooperation were drawn into preparing the budget request. In the case of the MFA, the Economic Cooperation Bureau's Policy Division coordinated the demands of its divisions to fit the budget limits laid down by the Ministry's Accounting Division and by the MOF. The pressures of past budgets, ongoing policy and projects, and the interests of participants determined the size of the divisional request.

Grant Aid

The grant aid budget covered projects to be committed and budgeting itself was important in policy-making, in setting restraints in new projects and policy directions. Since, however, both the MOF and the MFA each requested some grant aid (and the MFA with JICA's help implemented most of both), they had to consult. The MFA took into account the impact of MOF items on its own (and JICA's) staff resources while at the same time trying to increase its project grant aid allotment. Like any requesting ministry, the MFA had to be extremely sensitive to the balance between its manpower and the size of its budget, and careful not to overestimate its ability to spend its budget in any one year. Only in giving to multilateral aid agencies has its record consistently been above 90 per cent disbursement.[12] Thus, grant aid to Mongolia of 5,000 million yen, which was originally proposed by an MFA regional bureau, was incorporated in the MOF budget request, not only because it was quasi-reparations (which had always been requested by the MOF) but also because the Economic Cooperation Bureau feared that the large amount for this aid would upset the balance within the total MFA budget. The first instalment of 1,516 million yen was appropriated in 1977. This helpful arrangement permitted the Second Economic Cooperation Division to claim the projects it originally desired, free from the imposition of that type of politically motivated policy item. According to officials of the division, its budget was small enough without having to be defended against inroads by other MFA bureaus, but the rapid rise in the grant budget in 1978 and 1979 will complicate the Second Economic Cooperation Division's attempts to keep control over what items will be included in the request.

Multilateral Aid

The direct policy impact of the budget was strong also for multilateral aid, although it affected United Nations aid more than aid to international financial institutions, which was always given multiyear disbursement through the MOF budget. Payments in bonds were transferred to a special account. Multilateral aid brought in another bureau to MFA budgeting, the United Nations Bureau, which requested and implemented United Nations assistance budgets. Because of their size, large subscriptions to international agencies were often not finalised until late in the budget timetable. Thus, an appropriation in bonds of 80,000 million yen for Japan's share in the replenishment of the International Development Association was included in the 1977 budget only after a decision in late January, because of MOF hesitation in the face of delays in the United States Congress's authorising of American funding. The MOF was uncertain about putting legislation to the Diet approving increased funds before international agreement on cost-sharing was reached. The MFA was opposed to postponement, mainly on the rather parochial grounds that the IDA subscription represented a full 14 per cent of the fiscal 1977 ODA budget, removal of which would have sent Japan's budgeted ODA even lower than was expected.[13]

Technical Cooperation

Technical cooperation budgets were designed mainly to meet JICA needs, and here the MFA had to discuss (often reluctantly) its own requests with divisions of other ministries. Article 42 of the JICA Law required MITI, the MAFF, the MOF and the MFA to consult over the JICA budget but, in fact, MITI and the MFA were happy to operate fairly independently. JICA drafted requests in June which were based on its own forecasts of the following year's disbursement which had been agreed on in meetings between the relevant MFA/MITI/MAFF divisions and corresponding JICA departments. The MFA Economic Cooperation Bureau's Policy Division had the power (as the bureau's budgeting division) to express an opinion on the JICA budget, but usually the responsible MFA division directed JICA. Both MITI and the MFA drew up plans for regional and country distribution of technical assistance and used these to prepare their separate budget requests for technical aid programmes. JICA was able to discuss with ministries the practicality of these plans and its view was respected, provided that agency officials did not press too hard; MFA officers never hesitated to point out JICA's officially subordinate status.[14] In fact JICA (like the OECF) developed a deal of informal influence as the specialist agency, and was prepared to use it, even though the rules of the bureaucratic pecking order had to be seen to be observed.

In respect of agricultural aid, the MFA had to consult with the MAFF both directly (in the case of JICA loan funds) and indirectly, through the JICA agricultural cooperation departments (for ordinary technical assistance). Since the MFA actually requested the agricultural portion of the JICA budget (which did not occur for the mining and industry portion requested by MITI), the MFA was able to assert some influence over agricultural development policy. It was careful, however, to secure MAFF agreement on budgets, for influence was balanced; the MAFF in the end provided the main technical staff for agricultural projects and controlled the agricultural departments of JICA. MAFF officials naturally emphasised that their opinion predominated if problems arose about the content of requests, and while they could be 'direct and positive' in their approach, they were not always effective in forcing adjustments since, ultimately, the MFA made the request. Future cooperation and budgets required MAFF support, however, and the Economic Cooperation Bureau was mindful of its own political weaknesses. On the other hand, it was ready to play off the MAFF against MITI in any three-cornered argument, which often developed about the size of the budget for JICA loan finance.[15]

This haphazard and rather tense process of MFA-MAFF coordination was not followed for the JICA mining and industry-related budget, almost all of which was derived from MITI-commissioned funds (*itakuhi*). It was requested independently of other ministries' aid budgets by MITI's Economic Cooperation Department, after talks in mid-to-late June with the Economic Cooperation Bureau of the MFA on levels of request. There was no exchange of details. The Technical Cooperation Division of MITI drew up mining and industry technical cooperation policy[16] and, after reviewing this with JICA, incorporated the amount in its own budget application.

Once interministerial differences had been settled, the JICA request was included in the technical cooperation section of the Economic Cooperation Bureau's budget request, which usually totalled about 130 per cent of the previous year's allocation. The First Technical Cooperation Division and Policy Division fitted it, after some adjustments, into the Bureau budget, which faced its first test at meetings between the MFA Accounting Division and Directors of the Policy and General Affairs Divisions of the MFA bureaus in early July.

Budget requests for technical aid of all sorts were highly specific. Those for feasibility studies to be undertaken by JICA departments, for example, cited proposed projects by name, although these lists could be altered or shuffled about once the budget was finalised, as long as no increase in expenditure was necessary. Likewise, intended project-base technical cooperation had to be clearly designated, but budgets for trainees and specialists were less restrictive, having only to indicate proposed numbers of persons.

The Accounting Division then attempted to organise the many requests into the ministry budget application. This, again, was a detailed task involving individual programmes and expenditures weighed by directors whose influence was, of course, as dependent on their own personality and negotiating strength as on the intrinsic importance of their bureau's budget. These amendments were also shaped by ministry budget policy drawn up, in the case of the MFA in 1976, by the Director-General and General Affairs Councillor of the Minister's Secretariat and the Accounting Division Director, and issued in mid-July. Economic cooperation always featured in MFA budget policy because it occupied about 47 per cent of the whole ministry budget. In contrast, aid was only a minor element of MITI's expansive budgetary priorities.[17]

Government Loans

Preparation of the budget for government loans (for moneys to be disbursed in the following year and for some uncommitted disbursements such as commodity aid) involved an altogether different set of participants, principally the OECF, the Eximbank, the EPA and bureaus of the MOF; actual budget preparations for loans were in fact largely divorced from commitments in a way which grant aid was not. Loans budgeting was not entirely confined to the ministries as with other aid, mainly because funds for the OECF were not allocated to a ministry and then to an implementing body — as the JICA appropriation was — but were placed nominally in the MOF aid budget and drawn on directly by the OECF.[18] In formal terms, then, MITI and the MFA could not participate in loans budgeting, although they were members of the important four-ministry committee which approved government loans commitments and thereby directly affected levels of future disbursement and, consequently, budgetary allocation. The experience of the OECF over the middle years of the 1970s clearly revealed how a reduced level of commitment helped to restrain budgetary growth. MITI and the MFA were also informally involved in low-level discussions between junior ministry officers and OECF/Eximbank loans departments about individual projects and their effect on the budget.

As always, statutory procedures were far simpler than reality; the relationship between the agencies and their supervising ministries was familiar but by no means one-sided. The OECF presented its official request to the EPA, its supervising body, which passed it on to the MOF. The size of the request was based on several factors: the results of exploratory EPA discussions with the MOF on budget possibilities and EPA advice to the OECF on a suitable request limit; the level of commitments on new projects and expected commitments; and, importantly, the progress of disbursement to ongoing projects. This last was one of the main items of information on which the Budget Bureau assessed OECF requests, for substantial undisbursed funds implied blockages in the expenditure process which new budgets would not remove and could possibly worsen. Indeed, only the OECF was really aware of its exact cash flow, and it provided a monthly statement to the MOF via the EPA in order to draw on General Revenue. Continuing loan commitments prevented the MOF, or even the EPA, from being able properly to monitor the progress of disbursement. The OECF's record in spending its budget was far from outstanding, and fluctuated greatly from year to year. From 1973 to 1976, its rate of disbursement was 57.5 per cent, 71.1 per cent, 61.5 per cent and 66.6 per cent, due to several problems at both ends of the donor-recipient relationship. Surplus funds carried over into the following fiscal year comprised 20-30 per cent of total OECF budgets between 1973 and 1977. Under the aid-doubling plan, disbursement improved in 1978, 88.3 per cent of the budget for direct loans being spent.[19]

Once the OECF and the EPA had agreed on the request size, the OECF divided that sum into the amount to be drawn from the General Account and the borrowing limit it wanted from the Trust Fund, ensuring that as far as possible the cumulative balance of each remained in a ratio of 1:1 (from 1979 3:1). Requests were sent to the Budget Bureau and to the Financial Bureau respectively and the final balance was left to these two bureaus to settle, although disagreements were common, as observed above. Conflict between the two about the proportion of amounts from the General Account and the Trust Fund to be budgeted to the OECF appeared also in respect of the Eximbank budget. The MOF was not one ministry, it seemed, but many.

Differences were not all-pervasive however; relations between the Budget Bureau and the International Finance Bureau concerning the OECF budget were two-sided. Before making a request, the OECF discussed with the International Finance Bureau's Overseas Public Investment Division the structure and size of the request and the progress on projects included in it. It did this to provide the ministry, through its specialist division, with detailed knowledge of the application, and advice from the division assisted the Budget Bureau when making a final allocation. The International Finance Bureau thus advised both parties, acting as mediator. On occasions, as has been noted before, the division appeared to have a split, rather than a fully integrated personality, on the one hand advocating the OECF view to the Budget Bureau and, on the other, modiying the OECF request by arguments for restraint.

Before formal budget requests were made by the end of August, ministry submissions were discussed with appropriate committees of the LDP's PARC, both before and after the official Ministry Budget Conference had ratified the request. Officials met with the Special Committee on Overseas Economic Cooperation and with the Foreign Affairs Committee (for the MFA), the Commerce and Industry Committee (for MITI) and so on. In the case of the MFA, briefings were officially undertaken through the General Affairs Division of the Minister's Secretariat, not through the Accounting Division, and were said to be general explanations which altered few items. The earlier talks, however, which bureau directors attended, were an opportunity for the ministry to put its budget submission to the politicians, to seek their support and to allay complaints. As one former Accounting Division Director explained,[20] it was necessary that Diet Members' support be secured, their wishes catered for but their zeal for individual projects or items often restrained. The political limits within which ministry officials had to work were often narrow and the intricate relationships built up between ministries and Diet members, while often beneficial to both, involved delicate bonds of obligation.

The focal point for this budgetary politics was the director of the ministry's Accounting Division, and the last two holders of the position in the MFA were officers with considerable experience in the Economic Cooperation Bureau. Campbell likened finance officers to their American counterparts, hard-working, sincere men, well versed in their brief, but cunning and sensitive also, as Heclo and Wildavsky pointed out in relation to Britain.[21] The skill of the Principal Finance Officer there, they observed, lay in balancing internal demands with external conditions, understanding the market in budget politics; he had therefore to be something of the lion and the fox. In Japan, it was said, his skill was that of *haragei*, or 'belly talk', the politician's style of negotiation. It was his duty to know, despite his formal isolation from the political process, what politicians wanted and how far they could be manipulated or accommodated. While relationships of trust were vital, these had to be interpreted in the context of the demand for budgets. The finance officer had to act as the 'Budget Bureau' in his own ministry, and put his ministry's case firmly enough to Diet members to garner their support when required, but not so strongly that the trust built up between him and the MOF was jeopardised.[22] The formal complexities of personal relations in Japan, however, meant that duty, obligation and honour had to be seen to be fulfilled, although in a policy area such as aid, largely divorced from electoral politics, the demands of politicians on ministry budget officers were perhaps less onerous than on officers dealing with domestic issues.

Request and Response

The MOF was busiest at request time. Situated at the main intersection in Kasumigaseki, Tokyo's government district, the MOF had MITI before it, the EPA behind and the MFA and the MAFF across the road to the left. Ministry officials had to walk or be driven only a few minutes to join the queues in the Budget Bureau corridor waiting their turn to see the examiner and his staff and put their case. Requests for the economic cooperation budget were made independently by the ministries, to the Foreign Affairs-Economic Cooperation Desk of the MOF's Budget Bureau. Some of the ministries with small aid budgets requested directly through the examiner responsible for their own ministry, whereupon the aid portion was rerouted to the Economic Cooperation Desk.

This desk was a part of the larger section dealing with trade and foreign affairs under one of the Budget Bureau's twelve examiners. It was, like most offices in the old MOF building, an unprepossessing office, divided off from the larger room by a row of cupboards. The investigator in charge of the desk sat by the window and his half-dozen officers were positioned around desks in the middle of the room, while chairs for waiting ministry officials were placed along the walls.

The ministry request was compiled with the MOF directive on request limits in mind and the ministry's Accounting Division chopped and changed bureau requests to fit the broad ministry budget. Table 5.4 indicates that total ministry requests for three fiscal years 1975-7 kept very close to the limits imposed by the MOF, but that their economic cooperation request usually exceeded them by a wide margin. Economic cooperation budgets were not excluded from MOF guidelines during those years but ministries, especially those with aid budgets which were a minor portion of the total ministry budget (such as the EPA and Ministries of Education, Health and Welfare, Transport, Labour and Construction), could absorb the increases in the overall request. In the MFA, where economic cooperation took up about half

Table 5.4: Total Budget Request and Economic Cooperation Budget Request per Ministry, 1975-7: Request Divided by Previous Year's Allocation (Fiscal Year, Percentage)

	1975		1976		1977	
	Total	Economic Cooperation	Total	Economic Cooperation	Total	Economic Cooperation
EPA	na	433.4	na	120.9	na	115.2
MFA	127.3	132.2	115.0	118.9	113.5	129.5
MOF	124.1	99.6	113.7	117.3	112.2	136.8
Education	124.9	153.7	115.0	125.1	114.3	125.6
Welfare	129.7	156.0	122.8	121.4	121.6	105.8
MAF	85.9	175.6	113.5	131.5	113.1	205.0
MITI	123.8	158.3	114.4	137.0	111.5	126.2
Transport	123.9	132.1	114.8	113.0	114.1	253.6
Labour	124.8	161.7	114.8	163.1	113.3	191.6
Construction	124.7	212.3	114.5	164.0	114.6	134.3
Total	123.5	113.5	116.8	119.0	118.7	134.5
Request Limit	125.0		115.0		115.0	

na = not available.

Economic Cooperation requests represented requests for capital grants, technical assistance and multilateral aid. The EPA, Education, MAF and Construction give bilateral assistance only (technical assistance).

Source: *Nihon keizai shimbun* and MFA budget documents.

Table 5.5: Economic Cooperation Budget, 1975 and 1976: Sizes of Allocations and Requests (Fiscal Year, Percentage)

Ministry	1975 Bilateral Grants (including technical assistance)			1975 Multilateral Grants			1976 Bilateral Grants (including technical assistance)			1976 Multilateral Grants		
	75R/74A	75A/75R	75A/74A	75R/74A	75A/75R	75A/74A	76R/75A	76A/76R	76A/75A	76R/75A	76A/76R	76A/75A
EPA	433.4	25.4	110.3	–	–	–	120.9	88.5	107.0	–	–	–
MFA	130.2	87.7	114.2	136.1	91.0	123.8	126.5	88.9	112.5	104.3	95.2	99.3
MOF	95.2	93.5	89.0	101.2	87.8	89.0	50.3	96.9	48.7	142.4	94.9	135.1
Education	153.7	81.0	124.5	–	–	–	125.1	85.7	107.2	–	–	–
Welfare	181.8	77.5	140.9	155.3	99.5	154.6	135.5	66.7	90.3	121.0	100.0	121.0
MAF	175.6	72.4	127.1	–	–	–	131.5	72.9	95.8	–	–	–
MITI	156.2	75.5	117.9	253.9	52.5	133.3	137.6	79.9	110.0	113.2	95.5	108.1
Transport	136.9	48.3	66.2	115.8	259.1	300.0	111.6	112.5	125.6	114.0	96.9	110.5
Labour	181.6	56.8	103.2	119.0	100.0	119.0	134.9	87.9	118.6	215.9	97.3	210.1
Construction	212.3	54.3	115.4	–	–	–	164.0	87.8	144.0	–	–	–
Total	119.9	87.5	104.9	108.4	88.7	96.2	102.5	88.9	91.0	132.9	95.0	126.3

A = allocation, R = request.
EPA, Education, MAF and Construction gave no multilateral grants.

Source: MFA budget documents.

of the ministry budget, restraint in requests for other programmes was strictly enforced. Furthermore, there was no uniformity between the size of ministry aid request, which reinforces the evidence that officers in charge of aid budgeting in the ministries did not regularly communicate with each other about their requests, except in a perfunctory way when legally required.

The amounts requested by most ministries fluctuated, an indication that single programmes could determine the eventual scale of that request, especially when a ministry had only a few aid budget items. This is borne out by Table 5.5, which shows, for fiscal years 1975 and 1976, the proportionate size of ministry requests and allocations, and the percentage increase in annual allocation in both bilateral and multilateral grants. It can be seen there that the scale of economic cooperation requests varied greatly between ministries, as did the MOF response (allocation/request); there was no simple relationship between a large request and a proportionately small allocation. Substantial requests (such as those from the Health and Welfare or Labour Ministries for 1975) brought different percentage allocations (77.5 per cent to 56.8 per cent) and therefore increases of allocation in 1975 over 1974 diverged greatly (140.9 per cent to 103.2 per cent). Ministries with smaller aid budgets (Health and Welfare, MAF) could afford in bargaining terms to make requests in 1975 well over the 125 per cent limit and still receive an increase in their allocation much greater than that for the MFA. The MOF managed carefully to keep the MFA budget regular, but allowed individual programmes in, for example, the Transport budget, to swing about wildly: only 66 per cent allocation in 1975 for bilateral grants, but 125 per cent in 1976, more than was requested, while multilateral grants in 1975 received a 300 per cent allocation.

Table 5.6 goes further; from it we can see that when fixed expenditure items determined by non-budgetary factors were eliminated (by neglecting the MOF aid budget, which comprised in those years reparations and contributions to international financing agencies) most allocations in 1975 and 1976 remained within the range representing incremental increase (100 to 129 per cent of the previous year's budgeted amount[23]). A high proportion of items still lay outside this 30 per cent band, however, especially in 1976 when budget policy was tighter. This means that, although the previous year's allocation was an important determinant of the new budget, the MOF scrutinised individual programmes, both in the small and the large aid budgets, and approved increases where warranted. It could also disallow increases:

Table 5.6: Economic Cooperation Budget, 1975 and 1976: Allocation Divided by Previous Year's Allocation (Fiscal Year)

Allocation/ Previous Year's Allocation (%)	1975		1976	
	Items	%	Items	%
under 80	4	2.3	16	8.4
80-9	2	1.2	28	14.7
90-9	2	1.2	27	14.1
100-9	68	39.5	61	32.0
110-19	24	13.9	25	13.1
120-9	24	13.9	13	6.8
130-9	13	7.6	3	1.6
140-9	8	4.7	4	2.1
150-9	2	1.2	6	3.1
160-9	2	1.2	1	0.5
170-9	3	1.7	2	1.0
180-249	8	4.7	3	1.6
250 & over	12	6.9	2	1.0
Total	172	100.0	191	100.0

Excludes reparations and contributions to multilateral financial institutions (part of the MOF budget), but includes aid budget items for MFA, MITI, MAF, Education, EPA, Construction, Labour, Transport, Welfare, Justice and the Administrative Management Agency.

Budget requests were classified as headings, sub-headings, programmes and sub-programmes. Items counted included sub-programmes, programmes without sub-programmes, sub-headings without programmes etc., budgeted in 1974, 1975 and 1976 and those budgeted in just 1975 and 1976, making 172 items in 1975 and 191 items in 1976.

Source: MFA budget documents.

20 multilateral grant items, for example, did not rise. The outcome of the tussle over programmes was one indication of the seat of power in the aid bureaucracy — it certainly made the MOF a conspicuously involved member — but success in budget requests was obviously variable and could not be a reliable measure of generalised influence in aid policy. For example, the MFA's relative success in 1975 (87.7 per cent of request being approved and 29 of 100 items requested receiving an allocation of 130 per cent or more) was not matched in 1976 (88.9 per cent approval but all other ministries (except one) increasing their approval rate, and only 15 items receiving 130 per cent or more and 32 items in fact falling below 100 per cent). MITI items received a

similar fate in the two years.

If the economic cooperation budget were subject to mostly incremental increases, despite irregularity in some programmes as new items appeared and older ones dropped away, the MOF Budget Bureau probably approached ministry requests and explanations in much the same way as it did domestic budget programmes. The Budget Bureau was wary of the potential drain on budgetary reserves by aid budgets: thus the usefulness to the bureau of the FILP. The appeal by ministries for balance and fair treatment, their continued attempts to initiate planning, or their unsubtle references to international aid trends and the dire implications of a stagnant Japanese foreign aid effort (perhaps the most common weapon in the aid officer's budget strategy), could be easily countered by an astute Budget Examiner. The expansionary effect of the built-in 'contract' aspect of loans budgeting, for example, whereby advance commitments preceded budget allocation, was to some extent dampened by the implementation difficulties readily pointed to by the Budget Bureau; there was often good cause for caution. The same applied to requests for capital grants, the disbursement rate of which was normally low. The Budget Bureau's tendency to keep the discussion at the general level, avoiding talk of details and individual projects (in loans at least) where the ministries had an advantage, inhibited the give and take of negotiation on policy and let the bureau's superiority in the broader issues of budgeting prevail. Thus the Budget Bureau's arguments against the OECF's request often concerned levels of disbursement and the proposed scope of new commitments and their fiscal impact. The Budget Bureau recognised the need to increase aid budgets, admitted the ministries, but only in balance with other categories.

There were many MOF approaches to aid budget requests, some universal — balance, the strictures of broad ministry fiscal and budget policy and the implications of those for individual expenditures — others more immediate, in particular the skill of each Budget Examiner in grasping quickly the nature of aid programmes and their effect on the budget.[24] The need for a Budget Examiner to become rapidly familiar with aid was reinforced by the tendency for career officers (which the examiner (*shukeikan*) and his investigators always were) to move from post to post within the ministry every two years or so. Other ministries complained that by the time an investigator began to 'understand' aid, he was due to be transferred. Their complaints were all the more pertinent because the examiner and his staff exhibited the same 'double character' we saw in ministry Finance Officers. On the other hand, the MOF did not subscribe to the adage that 'a little learn-ing is a dangerous thing'; instead, they argued, it had its benefits: regular movement of officers was said to imbue successive examining staff with the 'amateur's eye' necessary for a commonsense assessment of aid and, in a wider context, ensured that officers remained un-committed to any one policy area. Yet, it was the investigators who had to argue for economic cooperation at the Budget Bureau Conference on each ministry's budget in late October or early November. These final meetings in the Budget Bureau to settle ministry allocations were, in the words of one investigator, 'a fight for the spoils'. Investigators had to know their subject and how far they could push for a fair share within the whole budget. The investigator therefore was one of the few individuals able substantially to affect the size of the aid budget, albeit at the margins, by and large all that was to be settled at that late stage.

Budget Bureau examinations were undertaken with examiners adopt-ing a typically strict attitude to foreign aid expenditure, an attitude based above all on domestic considerations: the effects of aid policy on the home economy and on the benefits accruing to the Japanese people, in other words the 'cooperative' and 'mutual benefit' aspects of economic cooperation. For this reason, it was argued that the Japanese people had to endorse aid-giving, and that the MOF's self-appointed task was to ensure that aid was economically sound and was kept within the limits imposed by fiscal restraint. Not only should aid be justified in strict (and, by implication, objective) economic terms, but it had also to be used efficiently. This austere standard was essentially no different from those applied to the other uses of public funds, but the diffuse, even divided, aid administration allowed the MOF view to prevail when the 'market in budget politics' might otherwise have prevented it. Furthermore, this attitude was expressed strongly and frequently in the year-round interministry negotiations on implement-ing the budget and on pending projects, and during extensive informal pre-budget discussions. Both ministries and the Budget Bureau accepted that the outlines of requests and the opposing Budget Bureau position were well known before the requests were actually made; constant communication fostered expectations of incremental change.

Mood-building

Success in achieving increased aid budgets was often dependent on the ministries' ability to build within the government a climate favourable to aid. In this vein, for example, overseas criticism of Japan's aid effort had a telling effect on aid budgeting. A severely critical response at the DAC Annual Examination of Japan in 1976 brought newspaper calls for a thorough restructuring of Japan's aid, including budgeting.[25] It also prompted agreement by the MFA and the MOF that every effort would be made to hold Japan's ODA at 0.24 per cent of GNP in 1977 by increasing grant and multilateral aid budgets.[26] They were external and rather *ad hoc* forces, though; the Policy Division of the MFA Economic Cooperation Bureau would try to direct more immediate pressures onto the MOF to improve the chances for favourable allocations to foreign aid. From its own intelligence work the Policy Division could build up a complete picture of the aid requests of the ministries and from that it could judge where and how best to lobby. The Policy Division Director himself often made the difference between a successful and an unsuccessful campaign, since he was most aware of trends in policy and was the bureau contact with the political and business world. He worked closely with those politicians interested in more progressive aid policies, one of whom was the late Minato Tetsurō. Indeed, a paper written and published in Minato's name in a respected development journal, which argued strongly for the implementation of a long-term budget framework for aid, was originally drafted by the Policy Division.[27] The MFA also sponsored a public relations campaign to raise public awareness of aid. The 1978 MFA aid report, much vaunted in the press as a rival to the MITI volume, was one product of this.[28]

The MFA also solicited the support of domestic client groups in budgeting, and preparatory informal negotiation and 'mood-building' (*nemawashi*) were not neglected. In 1976, for example, its Policy Division seemed pleased with its tactics in this respect. Not only had it encouraged September statements by the Advisory Council and the *Keidanren* on the need for a more positive government response to aid policy,[29] but had also worked with reporters from the *Nihon keizai shimbun* and *Asahi shimbun* attached to the MFA to publicise Japan's low aid commitment and loan disbursement, in an effort to prevent the MOF using these figures as a pretext for cutting budgets.[30] The resulting articles brought a sharp unpublished reaction from the MOF and an official response in a counter-article.[31]

Other ministries actively mobilised domestic client support, or rather used the momentum initiated by clients to promote budget requests in spite of staunch MOF reservations. One clear example was the approval in the 1977 budget of a scheme for bond insurance, part of a MITI campaign to step up exports of Japanese plant and equipment. The MITI budget request for 1977 was built around the need to increase exports[32] — a somewhat misplaced goal in view of the size of the expanding current account surplus — although MOF policy was slanted towards public works expenditure and away from export drives which it quite sensibly saw as causing adverse reaction overseas.[33] This policy included restrictions on the growth of allocations to the Export-Import Bank.[34] The bond insurance scheme was first proposed in 1975 because of Japanese failures to secure contracts on large scale projects. It was heavily backed by large construction companies (notably Kashima, which was able to go ahead with studies of the Nigerian rapid transit railway once the scheme was approved[35]), overseas trading companies and consultants, the Aid Advisory Council, *Keidanren* and by numerous politicians. It was finally agreed to in appeals talks between the Minister of International Trade and Industry, Tanaka Tatsuo, and the Minister of Finance, Bō Hideo, on 19 January 1977.[36]

Before the appeals negotiations, the LDP was also active, at all three levels which Campbell describes in his book on Japanese budgeting: rank-and-file members, unofficial groups and official party organs. After hearings with ministries before requests, little happened over the summer. Informal talks by ministries and agencies with LDP aid-related members, such as the Chairman of the Special Committee on Overseas Economic Cooperation (in 1978, Noda Uichi), were held before and after requests,[37] but it was not until the Diet reconvened in September that the party became fully occupied with the budget. There was a chance for individual members to use the Diet forum for their contribution to the aid debate,[38] but total party participation did not eventuate until after the MOF Draft was completed. In any case, the LDP's Budget Compilation Programme, released shortly before the MOF Draft, was a vague and, in relation to economic cooperation, quite useless document in any policy or budgeting sense.[39]

The Final Phase

The MOF Draft was shown to the ministries in December or early January, after which about ten days of highly political renegotiations took place. The appearance in the newspapers of the MOF Draft was the first occasion on which the economic cooperation budget became publicly known as a single policy expenditure, since no one ministry

Table 5.7: Budget General Account: MOF Draft and Government Draft Allocations for Various Categories, 1965-77 (Fiscal Year, Hundred Million Yen)

		1965	1966	1967	1968	1969	1970	1971	1972	1973	1974	1975	1976	1977
Social Security	MOF	4910.1	5989.0	6909.9	8122.0	9448.1	11252.4	13386.2	16338.5	21056.7	28812.0	38476.3	47355.8	56289.5
	Govt	5164.2	6217.1	7181.9	8156.6	9469.6	11371.1	13440.8	16414.7	21145.4	28906.8	39269.2	48076.3	56580.5
	Increase	254.1	228.1	272.0	34.6	21.5	118.7	54.6	76.2	88.7	94.8	792.9	720.5	291.0
	% Change	5.2	3.8	3.9	0.4	0.2	1.1	0.4	0.5	0.4	0.3	2.1	1.5	0.5
Education & Science	MOF	4634.1	5303.6	6000.3	6943.9	7982.6	9129.2	10612.5	12858.9	15266.0	19154.5	25983.5	29786.8	33704.1
	Govt	4757.4	5432.7	6246.1	7024.5	8057.4	9256.4	10788.8	13043.8	15702.3	19632.8	26400.7	30292.0	34297.4
	Increase	123.3	129.1	245.8	80.6	74.8	127.2	176.3	184.9	436.3	478.3	417.2	505.2	593.3
	% Change	2.7	2.4	4.1	1.2	0.9	1.4	1.7	1.4	2.9	2.5	1.6	1.7	1.8
Defence	MOF	2975.2	3366.9	3736.4	4220.7	4838.1	5695.2	6709.3	7998.4	9354.6	10930.2	13273.2	15123.5	16906.1
	Govt	3104.2	3406.5	3809.0	4220.8	4838.1	5695.2	6709.0	8030.0	9354.6	10930.2	13273.2	15123.5	16906.1
	Increase	129.0	39.6	72.6	0.1	0.0	0.0	-0.3	31.6	0.0	0.0	0.0	0.0	0.0
	% Change	4.3	1.2	1.9	0.0	0.0	0.0	0.0	0.4	0.0	0.0	0.0	0.0	0.0
Public Works	MOF	6587.7	8378.0	9615.2	10485.6	10934.8	13678.8	15979.0	20921.0	27464.7	28156.1	28864.1	34585.0	42010.5
	Govt	6926.7	8804.0	10005.0	10700.6	12063.6	14098.8	16655.9	21484.7	28407.7	28407.1	29095.2	35272.5	42810.5
	Increase	339.0	426.0	389.8	215.0	1128.8	420.0	676.0	563.7	943.0	251.0	231.1	687.5	800.0
	% Change	11.4	5.1	4.1	2.1	10.3	3.1	4.2	2.7	3.4	0.9	0.8	2.0	1.9
Trade Promotion & Economic Cooperation	MOF	107.5	188.2	301.9	460.3	637.5	899.0	965.5	1082.0	1210.4	1534.8	1677.4	1811.0	2063.9
	Govt	129.3	282.4	365.4	481.5	835.4	909.1	1011.0	1152.2	1287.6	1659.8	1767.1	1830.7	2108.6
	Increase	22.8	94.2	63.5	21.2	198.9	10.1	145.5	170.2	77.1	125.0	89.7	19.7	44.7
	% Change	21.2	50.1	21.0	4.7	31.2	1.1	15.1	15.7	6.4	8.1	5.3	1.1	2.2
Total MOF and Govt*		36580.8	43142.7	49509.1	58186.0	67395.7	79497.6	94143.3	114704.7	142840.7	170994.3	212888.0	242960.1	285142.7

*The total size of both MOF and government budgets (General Account) was the same.

Source: *Nihon keizai shimbun*, various issues.

requested all or most of the aid budget. The total economic cooperation budget was always increased during appeals negotiations. Whereas rises in other General Account categories were consistently small (with only Public Works allocations being raised by over 5 per cent at any time), economic cooperation was allowed wide increases, although the variations flattened out in the 1970s (see Table 5.7). These extreme changes during the appeals negotiations suggested that attention was directed towards specific programmes within the economic cooperation budget since, as the economic cooperation budget increased in gross terms and as a percentage of the General Account (see Table 5.2), so changes due to appeals negotiations stabilised.

Campbell sees appeals negotiations to be essential to the Japanese microbudgeting process. They allow broad and very public participation in bargaining at the margin for a fixed 'pot', force ministries to decide their priorities and make the final allocation of policy expenditures amid overt political role-playing and symbolic posturing. Some economic cooperation officials referred to them as an operation stage-managed by the MOF to assess real ministry wishes. The division of appeals into four stages, involving division directors, bureau directors, vice-ministers and then ministers in separate talks, enabled the MOF to direct the play. As appeals went higher, available options became clearer and ministerial stances more or less obvious, while 'preserving face' and 'playing to the galleries' occupied more time. Most allocations were solved before vice-ministers were called upon and problems which then advanced to the ministerial talks were those in which considerable political investment had been made.

The same economic cooperation items appeared regularly in revival negotiations. Capital allocations to the OECF, Export-Import Bank and to the OTCA (now JICA) were settled before or at the vice-ministerial discussions, but the OECF amount was often not approved until ministerial negotiations. Table 5.8 shows that appeals negotiations increases for the OECF and the bank were always made from both the Trust Fund and General Account/Industrial Investment Account, but they were unusually small. Additions were frequently made to the MFA grant or technical assistance budget, or to emergency aid. That loan allocations came up each year, however, suggested that, while vital to aid expenditures, the OECF budget was very political, and perhaps beyond the authority of the MOF to control. The MOF could only seek to restrain the expectations of ministries and politicians and, as Table 5.8 reveals, it may have been successful in this. Appeals increases for both sources of OECF funds were not large.

Table 5.8: Budget Allocations for OECF and Export-Import Bank, 1965-77: MOF Draft and Government Draft (Hundred Million Yen)

| | OECF | | | | | | Export-Import Bank | | | | | |
| | General Account | | | Trust Fund | | | Industrial Investment Account | | | Trust Fund | | |
	MOF (a)	Govt (b)	b/a%	MOF (a)	Govt (b)	b/a%	MOF (a)	Govt (b)	b/a%	MOF (a)	Govt (b)	b/a%
1965	10	10	100.0	–	10	–	275	290	105.5	919	919	100.0
1966	60	75	125.0	60	75	125.0	350	370	105.7	1090	1150	105.5
1967	75	90	120.0	75	90	120.0	370	430	116.2	1750	1850	105.7
1968	60	60	100.0	60	200	333.3	460	480	104.3	2290	2150	93.9
1969	214	224	104.7	276	276	100.0	595	635	106.7	2860	2820	98.6
1970	280	290	103.6	292	310	102.7	740	760	102.7	2710	2730	100.7
1971	305	330	108.2	380	400	105.3	630	650	103.2	3075	3140	102.1
1972	380	420	110.5	540	610	113.0	600	630	105.0	3950	4200	106.3
1973	530	560	105.7	628	695	110.7	580	630	108.6	4815	4865	101.0
1974	600	650	108.3	720	770	106.9	450	600	133.3	5565	5565	100.0
1975	610	650	106.6	840	945	112.5	550	620	112.7	5565	5595	100.5
1976	na	755	na	953	978	102.6	620	670	108.1	6486	7036	108.5
1977	740	760	102.7	960	980	102.1	550	630	114.5	8150	8220	100.9

(a) = MOF Draft Budget, (b) = Government Draft Budget, na = not available.

Source: *Nihon keizai shimbun*, various issues.

Campbell concluded that the appeals negotiations were the budget period in which the influence of the Liberal Democratic Party was most obvious and direct — this was true for economic cooperation, although as a rule the articulation of political interests concerning aid was less noticeable in budgeting than at other points in the lengthy development of aid policies and programmes. The party participated formally when ministries lobbied the LDP Special Economic Cooperation Committee, and the PARC Divisions, to seek support for increases in programmes and items. These approaches were incorporated in party representations to the MOF, depending on the party's own priorities and on the likelihood of settlement early in appeals. The personal interests of politicians were sometimes involved, for example in the creation of the bond insurance scheme or in the representations made by individual Diet members in the course of pursuing private political ends,[40] although individual approaches were frowned upon by the party. The Prime Minister himself, or senior ministers, could urge general increases in the aid budget, which occurred in early 1977 when economic cooperation was taken to the final party-government talks as the newly incumbent Prime Minister Fukuda unsuccessfully tried to achieve an aid budget which represented a future commitment to 0.30 per cent of GNP. By that stage, however, there was little scope for changing the overall size of the aid allocation.

Does Aid Equal Budgeting?

The structure of the Japanese aid system affected budgeting in two distinct ways. Because there was no central aid agency or ministry in Japan, the MOF, especially its Budget Bureau, scrutinised ministry programmes and, because of the lack of comprehensive forward planning for aid, provided the only overall coordination. Differences between ministries regarding their own needs and poor coordination between their individual programmes meant that there was no advance determination of the total size of the annual aid appropriation. Aid items were spread throughout the budget under different votes and the original negotiations about these were carried out separately with the Budget Bureau by over a dozen agencies and ministries. The total aid budget was not known publicly until the MOF released its Draft Budget in December.

MOF control of a ministry's participation in aid imposed severe constraints on that ministry in determining its own aid priorities. The gross amount of aid commitments was decided on budgetary considerations, and annual budgeting tied levels of aid committed to levels of past commitments. While the budget allowed opportunities for the growth and development of aid, they were restricted, and incremental expansion of many ministry aid budgets was, in the long term, harder to control and harmful to policy coordination. Aid policy as a budget category was assessed more as a collection of individual programmes than in macrobudgeting terms and, because there were no specific or detailed country policies, bilateral aid relationships could not be separated from domestic Japanese fiscal problems.

In contrast, the dispersion of the aid system and the predominance of loan allocations within the aid budget weakened Budget Bureau and MOF control of the eventual size of the aid budget. Request compilation by ministries illustrated how budget deadlines concentrated ministry thinking and forced decisions at low levels of aid decision-making, balancing the strong influence of the Budget Bureau on the broader outlines of policy. The resources of the Budget Bureau were not large enough to provide coordination of all aid flows, even if it so desired. While the Budget Bureau surveyed the whole span of aid policy, it was a fleeting and limited view, in contrast to the detailed scrutiny of separate sections of MFA programmes required when the MFA Economic Cooperation Bureau prepared its request. The MOF shared in decision-making on loans within the four-ministry committee system, and budgeting became a segment of loans policy-making and a link in the continuing cycle of bilateral aid relationships. There was increased scope for politics to enter into budgeting, and political constraints on the MOF in aid-budgeting multiplied. The wider participation of the MOF in decisions on the allocation of aid (compared to systems with an independent aid agency) was countermanded by greater politicisation of aid policy.

The nature of aid-financing, predominantly commitments to long-running projects, encouraged incrementalism in aid-budgeting. The small size of the total aid budget, however, enhanced the impact of individual programmes on it, contributing to instability in budget increases, fluctuations in the shares of ministries and to changes in the share of aid in the General Account. Budget inertia restricted aid to some extent, for the natural stability in budgeting practices reinforced traditional MOF suspicion of foreign aid as a worthwhile government pursuit.

Being unable to extract aid from the single-year budgeting framework, despite a multiyear perspective introduced by the aid-doubling plan, constituted a fundamental barrier to a more flexible aid policy and one in which forward planning and commitments could benefit recipient countries. This inflexibility contributed to the quantitative growth of the aid programme rather than to improvement in its content. Ministries were played off against the MOF and against each other and, as a result, debate centred not so much on policy *per se* but on the merits of one programme against others. Individual programmes and projects were emphasised and ministries confronted the MOF on that basis. While budgeting provided a salutory pressure for ministries to review and justify their own programmes, it could not, by its very nature, coordinate different levels of aid to produce a cohesive and forward-looking policy. Budgeting looked to the past to order its short-term future.

Notes

1. OECD, Development Assistance Directorate, *The Management of Development Assistance: A summary of DAC countries' current practices*, 21 August 1975, pp. 13-19, and George Cunningham, *The Management of Aid Agencies: Donor structures and procedures for the administration of aid to developing countries* (London, Croom Helm, 1974), pp. 44-8. Nine DAC countries supplied ODA funds by appropriation through central government budgets, while eight others used a selection of additional sources: central banks, capital markets, amortisation receipts, special funds, national lottery, regional government funds. Unspent funds could be carried forward liberally in Canada, while a number of other members had carry-over procedures of varying flexibility. Multiyear planning was also practised by some governments, in the form of rolling estimates (Germany, United Kingdom) or specific targets (Belgium, Denmark, the Netherlands, Norway, Sweden).
2. Hugh Heclo and Aaron Wildavsky, *The Private Government of Public Money: Community and Policy inside British Politics* (Berkeley, University of California Press, 1974), pp. 343-60.
3. John Creighton Campbell, *Contemporary Japanese Budget Politics* (Berkeley, University of California Press, 1977).
4. Supplementary budgets were also passed in the autumn to adjust government employees' salaries and the rice price support scheme etc., and in recent years aid funds have been increasingly made available. In 1978 the supplementary aid budget was about 10 per cent of the standard budget allocation.
5. The main budget was divided into the General Account, for allocation from taxation revenue to major policy areas, and into Special Accounts for allocations for special government purposes. In 1975 there were 41 special accounts in operation. The General Account expenditure was presented in ten principal policy categories (social welfare, education, bonds, local finance, defence, public works, economic cooperation, small and medium industries, food management and pensions). Apart from these accounts, an important feature of the budget strategy was the annual Fiscal Investment and Loan Programme, which drew on several special accounts for allocations to government organs such as

hospitals, universities, corporations, funds, agencies and local government bodies. The chief component was borrowings from the Trust Fund (*shikin unyōbu shikin*), which derived its funds from postal savings, government health insurance contributions and government pension schemes. The Industrial Investment Special Account, which comprised moneys allotted from the General Account, was established in 1953 to finance industrial and export growth but in 1977 provided only limited capital, mainly to the Export-Import Bank (Eximbank).

The Overseas Economic Cooperation Fund (OECF) could accumulate borrowings from the Trust Fund up to an amount equal to the cumulative total of its capital and reserves (OECF Law, Article 29.3, amended in 1979 to a ratio of 3:1) and the Eximbank an amount up to four times (Article 18.3). Small amounts for emergency aid could also be drawn from the Emergencies Contingency Fund (*yobihi*), a portion of the General Account set aside for unforeseen (and at the time of Diet approval, unspecified) expenditures. It was rarely used following criticism of the government in the Diet in July 1967 after an amount was spent to supply part of a relief loan to Indonesia. The Japan Socialist Party demanded that large drawings from the fund be given Diet approval (see *Nihon keizai shimbun* (hereafter *Nikkei*), 14 July 1967). For general descriptions of budgeting, see Kōno Kazuyuki, *Yosan seido* (*The budget system*) (Tokyo, Gakuyō shobō, 1975), and Yoshino Yoshihiko, *Zusetsu nihon no zaisei: shōwa 50 nendohan* (*Japan's finances illustrated, 1975*) (Tokyo, Tōyō keizai shimpōsha, 1975), pp. 264-302.

6. The Special Reparations Account was established in 1956 by a law of the same name. Unspent funds could be carried over to the following fiscal year. Both this and the Special Account for the National Debt Consolidation Fund received allocations from the General Account. See Zaisei chōsakai, *Kuni no yosan 1965* (*The nation's budget 1965*) (Tokyo, Dōyū shobō, 1966), pp. 731 and 735, an official explanation of the budget published on behalf of the MOF.

7. There were three main types of carry-over: (a) ordinary (*meikyō kurikoshi*), for funds expected not to be fully disbursed in the financial year; (b) accidental (*jiko kurikoshi*), for funds of which the disbursement was unavoidably delayed; or (c) continual (*keizoku kurikoshi*), for funds expected to be disbursed over a number of years. Administrative procedures for carry-overs were extremely complex. See Harada Shūzō, *Kurikoshi kessan jimu hikkei* (*Procedures manual for carry-overs and audits*) (Tokyo, Ōkura zaimu kyōkai, 1975). The amounts carried over were usually quite large, especially for grant aid. Approval was mostly given automatically for *meikyō* and less frequently for *jiko*, when typhoons, political disturbances etc. held up payments. Budget papers for the years 1965-76 revealed requests for carry-overs from all ministries giving economic aid. Carry-over for loans was possible, but was an internal accounting problem for the OECF, not one which involved the MOF. For a brief description of the carry-over system, see Kōno, *Yosan seido*, pp. 170-4.

8. Campbell, *Contemporary Japanese Budget Politics*, p. 210.

9. Edgar C. Harrell, in 'Japan's Postwar Aid Policies', unpublished Ph D dissertation, Columbia University, 1973, Chs. 3 and 4, examined Japanese aid budgets of the 1960s and concluded that 'the most important determinants of the Japanese aid budget, whether ODA or ODA plus OOF, were the gross resources of the country (gross national product) and its trade balance. Aid budgets, on the average, tended to increase with the level of economic resources and with favorable trade balances' (p. 215). This conclusion is valid, although it should be tempered with consideration of the political aspects of budget-making.

10. This, as Campbell pointed out (*Contemporary Japanese Budget Politics*, pp. 211-17), was a typical budgeting tactic of ministries, although it was rarely respected by the MOF, except in the case of the Defence Build-up Plan, which

was useful in isolating controversial defence budgets from the annual budgeting process.

11. See Gaimushō keizai kyōryokukyoku, *52 nendo gaimushō shokan keizai kyōryokuhi* (Ministry of Foreign Affairs, Economic Cooperation Bureau, *Economic cooperation budget for the Ministry of Foreign Affairs for fiscal 1977*), 21 May 1976. The plan was a tame one, being based on an ODA target which the MFA would have preferred to be higher, and directed primarily towards maintaining the MFA aid budget.

12. *Gaimushō keizai kyōryokukyokuchō-hen, Keizai kyōryoku no genkyō to tembō: namboku mondai to kaihatsu enjo* (Ministry of Foreign Affairs, Economic Cooperation Bureau Director General (ed.), *Economic cooperation: present situation and prospects: the North-South problem and development assistance*) (Tokyo, Kokusai kyōryoku suishin kyōkai, 1978), pp. 355-60.

13. See *Nikkei*, 15 December 1976, and *Asahi shimbun* (hereafter *Asahi*), 21 and 25 January 1977.

14. The JICA budget was divided into two categories: economic cooperation and emigration. Economic cooperation was allotted most out of the total JICA budget and comprised three parts: government subsidy (*kōfukin*), a direct contribution through the MFA budget for technical assistance; capital (*shusshikin*) for JICA loans; and commissioned funds (*itakuhi*) which were direct grants to JICA from the MITI budget for technical assistance in mining and industry. Both the MFA and MITI received funds to pass on to JICA but in fact *kōfukin* and *itakuhi* were different kinds of subsidy. The former OTCA received *itakuhi* from both ministries but when JICA was established it was decided to give *kōfukin* instead. MITI, however, was adamant that its *itakuhi* would remain as such, for it allowed MITI much closer control of the eventual management of funds.

15. Loans were financed from JICA capital (*shusshikin*), a separate item which was included in the MFA budget. The MFA therefore had to consult MITI and the MAFF beforehand to obtain agreement on the size of the request, which covered proposed financing of projects in all sectors, including agriculture and mining and industry. It was not easy to reach consensus between the three ministries and JICA, and their differences did not end with the request. In 1976 the MFA had to request an amount much lower than other ministries wished because of substantial unspent funds (expenditure of only 7.7 per cent, 29.9 per cent and 18 per cent 1974-6). This caused considerable argument and reaction from ministry-connected Diet members, so that the request of 15,400 million yen was exceeded by an allocation of 17,200 million yen!

16. See Tsūshō sangyōshō tsūshōseisakukyoku keizai kyōryokubu gijutsu kyōryokuka, *Gijutsu kyōryoku kankei yosan no jūten* (Ministry of International Trade and Industry, International Trade Policy Bureau, Economic Cooperation Department, Technical Cooperation Division, *Features of the technical cooperation budget*), September 1976.

17. MITI conveniently provided a full account of ministry policy and its translation into budget requests in its ministry journal, *Tsūsan jānaru*.

18. The OECF prepared monthly cash flow reports which it sent to the MOF via the EPA in order to draw on available funds. General Account subscriptions were used first and then borrowings were made from the Trust Fund as necessary, up to the limit allowed. Analysis of loans budgeting is restricted to the OECF, the main source of ODA loan funds. Discussing the Eximbank budget would take the story well outside the ODA framework.

19. OECF documents.

20. Interview, 17 July 1976.

21. See Campbell, *Contemporary Japanese Budget Politics*, pp. 20-2, and

Heclo and Wildavsky, *The Private Government of Public Money*, pp. 120 ff.

22. A former ministry Finance Officer, in an interview on 17 July 1976, illustrated this by saying how he had to convince an LDP rural Diet member that his ministry could not seriously request an extra 500 million yen for a particular research institute, when Japan's contribution was then only 200 million yen.

23. Analysis of items for each ministry shows the same pattern. Out of 100 items in the 1975 MFA budget, for example, 67 were in the 100 to 129 per cent range, while in 1976 the figure was 65 out of 112.

24. Refer to Campbell, *Contemporary Japanese Budget Politics*, Ch. 3, and to Kojima Akira, 'Nihon no zaimu gyōsei' ('Japan's fiscal administration') in Tsuji Kiyoaki (ed.), *Gyōseigaku kōza 2: gyōsei no rekishi* (*Studies in administration 2: administrative history*) (Tokyo, Tōkyō daigaku shuppankai, 1976), pp. 161-215 at pp. 189-92.

25. *Nikkei*, 11 July 1976.

26. *Nikkei*, 23 August 1976. This was not successful.

27. The paper is Minato Tetsurō, 'Shiron: keizai kyōryoku hakusho: 12 kōmoku teigenshō' ('Economic cooperation white paper, a private version: a summary of my 12-point proposal'), published in *Kokusai kaihatsu jānaru* (*International development journal*), 10 June 1976.

28. A glossy 70-page booklet, *Me de miru namboku mondai: hatten tojōkoku to watakushitachi* (*Looking at the North-South problem: the developing countries and us*), was published by the MFA in 1976. It explained the North-South problem and economic cooperation in everyday terms and was followed by another dealing specifically with technical cooperation.

29. Reported in *Nikkei*, 1 September 1976.

30. See stories in *Nikkei*, 20 December 1976, and *Asahi*, 28 December 1976. The *Asahi* story highlighted low loan disbursement rates and predicted worse ODA/GNP results in 1976, while *Nikkei* reported Western criticism of Japan's aid effort. The MFA official interviewed also claimed that the MFA had succeeded in getting a story of DAC criticism of Japan's aid performance at the top of the national news for the first time ever. This could not be confirmed. The word used to describe ministry cooperation with reporters was *kettaku* (collusion).

31. An unofficial response complained of the ministry's being 'got at', while the formal reply appeared in an article in the newspapers on 7 January 1977 (see, for example, *Mainichi shimbun*). In this article the MOF emphasised its commitment to increased aid.

32. See article in *Nikkei*, 8 September 1976.

33. *Nikkei*, 21 November 1976.

34. *Nikkei*, 23 December 1976.

35. *Nikkei*, 24 January 1977.

36. The LDP Special Committee on Economic Cooperation was solidly behind the scheme, which the Ministerial Committee on Overseas Economic Cooperation had debated over the course of a year or so. One MITI official interviewed (26 October 1976) said that Diet members close to construction firms, especially those involved in work on large social infrastructure projects overseas, were applying pressure to have the scheme set up. When MITI made the appropriate request in the budget, the Construction Ministry also requested funds to establish an Overseas Construction Project Guarantee Corporation, although it did not expect its request to receive a favourable response because of the strength of MITI's argument. The 1977 budget included an allocation to MITI for payments of up to 400,000 million yen in any one year. Premium was set at 0.1-0.3 per cent and the rate of protection at 70-90 per cent. See *Nikkei*, 20 January 1977.

37. Top OECF officials had budget talks with Tanaka Tatsuo on 29 July 1976.

38. See questions by Tanaka Tatsuo in the General Session of the House of

Representatives on 27 September 1976.

39. Item 11.2 of the LDP Programme reads: 'Strengthening of International Cooperation: because Japan's future prosperity, and the welfare and stability of the LDCs, depend on improved mutual relations, we shall increase bilateral grant cooperation and technical cooperation with these countries.' See *Nikkei*, 6 January 1977.

40. One LDP Diet member closely associated with aid and United Nations affairs boasted of his talks with the various ministers and with the Prime Minister, in interviews on 3 and 21 January 1977.

Part III:
The Politics of Aid Relationships

6 PROJECTS, SURVEYS AND CONSULTANTS

Projects were the 'building blocks' of Japanese aid policy and, as earlier chapters have suggested, provided aid administrators with a fixed standard for procedures and a tangible monument to officials' efforts. Projects were visible and measurable; given the problems of coordination faced by the bureaucracy and the complexity of procedures, project aid (which formed the bulk of Japanese bilateral assistance) gave a concrete form to requests and to the long process from identifying projects through to approving and implementing the aid programme.[1]

Assessment and the selection of projects, as the basis for decisions on aid-giving, were not made on rational economic criteria alone. The nature of aid requests and their associated projects were dependent on the entire bilateral aid relationship, and — as this and the following chapter will detail — the very way in which projects were brought into the policy-making process of itself restricted options. Formally, however, project identification and feasibility studies (or development surveys) were the usual method by which projects were made ready for official decision on loans or grants.

Development surveys were essential to Japanese aid policy-making. They provided the hard information on which aid projects were assessed for 'aid worthiness', and were the chief source of general information on the detailed economic needs of recipient countries. An accumulation of survey reports provided a rich body of intelligence, if used for that purpose. In addition, they effectively filtered aid proposals and relieved officials of much of the responsibility of sorting requests. In this, they served to justify future aid flows and influenced the direction and size of aid to individual recipients. Project feasibility was a gateway to aid beyond the immediate project.

Surveys did not, however, make all decisions easier for bureaucrats. Indeed, the system's limitations — budgets, manpower, regulations — meant that to develop a manageable aid programme preferred proposals needed to be selected before studies were made. Feasibility became a more sophisticated criterion for sorting an already restricted list of potential aid projects. Consulting engineers (or, more commonly, 'consultants') were therefore also essential to policy-making, since narrowing the number of projects to be surveyed was in itself a sensitive policy choice. The present chapter explores this important, largely unnoticed, area of the aid process, while Chapter 7 considers the effects of bilateral relationships and Chapter 8 assesses the role of the funding agencies in reinforcing aid patterns.

Types of Surveys

The development survey was the most common form of project assessment. The Japanese Government required proposals from prospective recipients for project loans to be accompanied by a completed feasibility study, and preparation to that point was regarded as the recipient's responsibility, although the Japanese Government still assisted by financing feasibility studies within its technical aid programme. This work was carried out under the supervision of the Japan International Cooperation Agency (JICA), and could involve teams of officials, or companies contracted to do the work, spending weeks or months (even years if the project were a big one) in the field, often under extremely difficult and trying conditions.

The aid programme incorporated several kinds of official surveys, grouped generally in the three main categories of project finding and identification, feasibility study, and project design. Funds for this technical aid were drawn from the MFA and MITI budgets and most were passed on to JICA to be used (see Chapters 4 and 5). Tables 6.1 and 6.2 show the budgets allocated for survey work by the Overseas Technical Cooperation Agency (OTCA) (1962-73) and by JICA (1974-6), and the number of teams sent overseas to conduct surveys.[2] Funds from both MFA and MITI budgets rose substantially over the period (from 174 million yen to 6,738 million yen), but this was not matched by increases in the number of survey teams actually sent (from 19 to 113, see Table 6.2). Yet, there was a greater diversity of types of surveys shown in the official figures after 1970 and some (such as resources and mapping) experienced very rapid growth in the budget allocated to them. Although figures published by JICA provide data only to 1976, more recent budgets indicate a continued commitment to surveys, with increases of 23 per cent in the JICA budget for surveys in 1977 and again in 1978. The positive Japanese response to a more diversified role for surveys is shown by the special allocations for surveys for large-scale projects and for 'integrated regional development plans', and the initiation in 1978 of 'engineering loans', or loans for OECF surveys of candidate projects.

The geographical distribution of surveys carried out (Table 6.3) reflected more generally the spread of Japanese aid itself. Of the total number of 612 teams sent between 1962 and 1976, 360 (or 58.8 per cent) went to Asian countries, with an even balance of the remainder between South America and Middle East/Africa. The weight of the latter two regions in surveys increased later in the period, although it was not until after 1973 that it broke above one-third of the total and by 1976 had reached 46 per cent. Similar patterns showed in the spread of funds for surveys (Table 6.4), although the emphasis was slightly more biased towards the Asian region until 1973. None the less, by 1976 Asian countries received only 42 per cent of survey funds, while Africa and Oceania had begun to receive much-boosted allocations.

As total aid flows favoured certain recipient countries, so did development surveys. Indonesia was host to the most surveys, 96 between 1962 and 1976, followed by the Philippines (49) and Thailand (45). There was, however, no direct link between surveys completed and aid approved, for the time-lag in making a loan decision could be substantial and while feasibility was necessary for Japanese Government approval of a request for project aid, it was not a sufficient condition. Nevertheless, the concentration of project aid on a few recipients — to be detailed in Chapter 7, but especially Indonesia — was matched by the concentration of surveys. One was impossible without the other and each 'fed off the other'.

Because different types of surveys were needed at successive stages of a project, completing one project could require three or four surveys, often through different agencies. Even though identifying projects could be informal — often via channels well outside the regular aid bureaucracy, as will be explained — the survey could help formalise and organise a more ready assessment of the project. Surveys themselves could provide a vital link in the ongoing donor-recipient relationship beyond the span of individual projects. By their nature, surveys involved the analysis of conditions outside the immediate project site; they laid the foundation not only for later aid to the same project, but also to other projects. Surveys, and those who carried them out, were a motive force in the accumulation of aid ties with individual countries.

JICA and Surveys

Official funds for surveys were incorporated in the JICA budget

Table 6.1: Trends in Development Survey Budget, 1962-76 (Million Yen)

	1962	1963	1964	1965	1966	1967	1968	1969	1970	1971	1972	1973	1974	1975	1976
General	129	145	145	155	274	195	219	248	297	462	737	1047	1377	2085	3511
Pre-feasibility									3	26	31	30	52	72	74
Feasibility									291	381	470	557	851	1273	2194
Long-term											53	116	110	117	107
After-care									3	11	12	12	12	13	13
Mapping										44	171	332	352	382	609
Integrated Development Agriculture, Forestry & Fisheries														36	41
														192	436
Detailed Design							100	65	150	150	150	150	460	460	391
Special Projects													184	92	156
Development Planning	45	65	65	75	80	95	88	118	138	138	223	351	557	955	1434
Resources Survey									165	337	509	679	859	1209	1223
Other										7	9	124	297	219	60
Total	174	210	210	230	354	290	407	431	750	1094	1628	2351	3734	5020	6738

Source: Kokusai kyōryoku jigyōdan, *Kokusai kyōryoku jigyōdan nenpō* (Japan International Cooperation Agency, *Japan International Cooperation Agency Yearbook*), 1977, p. 106.

Table 6.2: Number of Development Survey Teams Sent, 1962-76, by Type

	1962	1963	1964	1965	1966	1967	1968	1969	1970	1971	1972	1973	1974	1975	1976
1. MFA Budget															
Feasibility	14	12	12	12	15	13	13	16	15	14	17	19	19	17	25
Pre-feasibility									1	9	13	11	10	15	10
Detailed Design							4	2	3	1	3	3	2	1	1
Agriculture														7	24
Other									1	3	3	7	9	8	18
2. MITI Budget															
Development Planning	4	9	8	7	6	8	8	8	10	8	10	12	18	24	24
Resources									1	3	5	6	8	9	11
3. Other	1				1						1				
Total	19	21	20	19	22	21	25	26	31	38	52	58	66	81	113

Source: As for Table 6.1.

Table 6.3: Number of Development Survey Teams Sent, 1962-76, by Region

	1962	1963	1964	1965	1966	1967	1968	1969	1970	1971	1972	1973	1974	1975	1976	Total
Asia	13	10	11	13	15	17	18	20	24	22	32	37	31	43	54	360
Middle East and Africa	1	5	4	2	2	2	4	4	5	8	8	11	18	20	31	126
Central and South America	5	6	5	4	5	2	3	2	2	6	10	9	15	17	21	112
Other										2	1	1	2	1	7	14
Total	19	21	20	19	22	21	25	26	31	38	52	58	66	81	113	612

Source: *Kokusai kyōryoku jigyōdan nenpō*, 1977, pp. 108-10.

Table 6.4: **Geographical Distribution of Development Survey Expenditure, 1954-76 (Million Yen)**

		1954-67	1968	1969	1970	1971	1972	1973	1974	1975	1976	Total
Asia	1	345.5	121.1	161.0	290.0	292.3	644.1	912.7	1270.9	1366.1	1644.3	7921.8
	2	143.4	55.4	61.2	216.1	153.9	438.3	490.7	418.6	624.1	816.0	3417.6
Middle East	1	86.2	–	14.8	3.0	10.5	31.1	7.6	30.9	165.9	310.6	660.6
	2	6.1	20.7	7.1	–	13.3	22.7	47.3	115.5	198.6	592.6	1023.9
Africa	1	32.1	22.7	40.5	62.9	89.5	47.8	279.8	340.6	540.7	624.2	2080.7
	2	42.7	–	11.2	28.9	12.3	3.9	98.8	212.0	106.5	86.4	601.6
Central and South America	1	70.7	8.0	11.6	12.2	78.4	98.4	52.7	128.0	360.0	415.4	1257.0
	2	149.5	20.7	23.3	8.2	94.6	187.1	203.6	245.2	379.0	1018.4	2329.6
Oceania	1	–	–	–	–	14.2	–	–	15.0	27.0	79.2	107.9
	2	–	–	–	–	–	–	–	68.2	278.9	319.4	666.5
Other	1	–	–	–	–	5.0	–	52.2	3.2	14.4	20.1	53.8
	2	–	–	–	–	–	3.3	–	–	–	–	3.3
Total	1	534.5	151.7	227.9	368.2	490.0	890.1	1305.1	1788.1	2447.1	3093.8	12126.2
	2	341.7	96.8	102.7	253.2	274.1	654.3	840.3	1059.6	1587.1	2832.8	8042.5

1 = development surveys of different types, from MFA budget (Item 1, Table 6-2).
2 = development planning surveys and resources development surveys, from MITI budget (Item 2, Table 6-2).

Source: *Kokusai kyōryoku iigyōdan nenpō*, 1977, Statistical Annexe, Tables 3 and 13, pp. 399-401 and 434-5.

following the MFA and MITI budget requests. In 1977 the development survey vote was 5,236 million yen and, from MITI, 3,147 million yen for overseas development planning surveys, representing some 20 per cent of the whole JICA budget for 1977. In 1978, however, this figure rose by 23 per cent to 10,352 million yen.[3]

JICA's development survey functions were managed by four departments, each (except for the Planning and Coordination Department) concerned only with its specified tasks and supervised by a different ministry. In coordinating the technical aid budget, the Technical Cooperation and Development Cooperation Divisions of the MFA's Economic Cooperation Bureau set guidelines, but translating these into concrete programmes was left to the operations departments of JICA. The Mining and Industry Planning and Survey Department came under the indirect control of MITI's Technical Cooperation Division and, to a lesser extent, the Resources and Energy Agency, by which it was commissioned to conduct resources development surveys.[4] The Agricultural and Forestry Planning and Survey Department worked closely with the International Cooperation Division of the MAFF's International Department, although formal contact was made via the MFA's Economic Cooperation Bureau. The Social Development Cooperation Department administered development work in other sectors, such as construction, transport, welfare and so on. While the department was responsible to the MFA, other ministries (Construction, Health and Welfare, Transport, Posts and Telecommunications) had a say in its management.

The size and scope of the year's programme for development surveys were determined in budget talks between JICA and ministry officials, leading to the budget request and the MOF response. JICA departments were able quietly to influence the structure of the request and its distribution by region, country and sector through informal and formal channels, usually through personal and working relationships between individual officers. The backlog of requests for surveys was often three times the limit of available resources, so the final selection was subject to considerable pressures. Surveys linked to loan-base projects had priority in talks with the First Economic Cooperation Division of the MFA Economic Cooperation Bureau. Some critics rather ungraciously dubbed the JICA survey budget *domburi kanjō*, or 'scrambled-egg accounting', where post-budget expenditure bore no relation to pre-budget requests. Certainly the names of surveys requested at budget time were only a preliminary ordering of priorities. The final sorting was done after the budget was decided in January and February, by the different JICA departments in consultation with ministries, and the basis for this was usually far from the apparently rational semi-programming leading up to budget request: survey policy was as dependent on the pressures of the bilateral aid relationship as any loan approval.

Administrative and economic arguments were not the only criteria for sifting survey proposals; the same systemic distortions appeared here as they did with other forms of aid. Indonesia, Thailand and similar well-placed recipients were traditionally given precedence and, in the Indonesian case, the project listings prepared for the meeting of the Inter-Governmental Group on Indonesia were referred to. Links to loan aid were weighed in the light of the availability of loan credits to particular countries: new loan recipients often had to wait between four and five years for the next credit, a delay which affected the timing of surveys. Project cost was another consideration, and recipients were said to be ranked according to allowable limits of expenditure on projects, presumably calculated by credit worthiness, economic prospects and the like.[5] Furthermore, since development surveys undertaken by technical assistance charted the way for capital aid in the years ahead, regional bureaus and other divisions in the MFA Economic Cooperation Bureau (or MITI) could be involved, depending on the country or the interests at stake. Pressures from companies likely to carry out surveys were also apparent, directed (as will become apparent) towards officers of the ministries and JICA.

Another factor was the type of survey envisaged. The three broad successive stages in project surveying — investigation, feasibility, construction design and supervision — were separate policy items, the success of one being a prerequisite for approval of the next. It was possible for all stages to be carried out by the same group of engineering consultants, but this would not always happen. Preliminary studies were made by teams of officials from JICA and related ministries, although project finding and identification were sometimes contracted out to advisory organisations like the International Development Center (IDC), Institute of Developing Economies, Japan Consulting Institute and the Engineering Consulting Firms' Association (ECFA). Surveys could be undertaken by groups of private firms under the sponsorship of the ECFA or by single companies themselves. The results of surveys conducted by such groups were, in principle, made available only to the commissioning agent but, in practice, were dispersed irregularly to other interested parties. IDC teams, for instance, could comprise specialists from government, business and academic circles.

It was usual for the later, technical stages of project assessment, such as feasibility and project design, to be contracted to engineering consultancy firms. This practice was a result of the traditionally close association of the consulting industry with aid projects and of the fact that in the late 1970s JICA still had insufficient technical staff or experience to itself carry out surveys on any large scale, in spite of the conviction of many officials that JICA should develop this capability. Engineering consultants were part of a small but growing industry which was dependent on government assistance and on work derived from government-sponsored projects. At the same time, they filled an important gap in the life of projects at a stage where problems were numerous and could easily lead to the breakdown of the aid process. Rondinelli suggests that these difficulties were: differences in perceptions and goals among funding agencies, recipients, technical experts and others; insufficient appreciation of local conditions; and inadequate preparation and design skills.[6] Consultants bridged stages in project development; they could offset these types of problems and compensate for the lack of expertise and local knowledge in the Japanese domestic administration. Interdependence with the decision-making process was their watchword, however, since aid policy was, for some companies, their own *raison d'être*.

The Engineering Consultants' Industry

In contrast to the West, where engineering consulting services developed first in the nineteenth century, Japan's industry was still young in the 1970s and grew after the Second World War under the influence of a few energetic and determined men.[7] While firms depended on an upsurge in the domestic economy for their survival, the extension of the Japanese economic presence into Asia was assisted by the industry; so too was the expansion of exports of Japanese heavy manufactures.[8] The early years after the war saw the establishment of a number of consulting firms, largely independent of existing business interests, and by 1951 Nippon Kōei (established June 1946) and Pacific Consultants (September 1951), later two of the largest and most successful firms, were in existence. With the increasing expansion of the Japanese economy after the mid-1950s, the consulting industry grew rapidly, and between 1954 and 1965, 31 firms were set up. In 1964 the ECFA was established. Japanese reparations contracts were responsible for much of the early growth in the overseas consulting industry, and its further development in the 1960s was closely associated with swelling Japanese Government aid flows and with contracts resulting from assistance by multilateral organisations to Asian countries.[9] Southeast Asian economic development provided consultants with their most valuable market. This experience was typified by the case of Nippon Kōei, the oldest and largest civil engineering consulting firm in Japan.

Nippon Kōei and Foreign Aid

The roots of the Nippon Kōei Company lie in prewar days, when Kubota Yutaka founded the Korea Power Company (*Chōsen denryoku kabushiki kaisha*) and the Yalu River Hydro-Electric Power Company (*Chōsen-manshū ōryokkō suiryoku hatsuden kabushiki kaisha*).[10] His companies completed many major development projects in Korea and Manchuria during the latter days of Japanese rule, particularly dams and water resources projects. Before and during the Second World War, their work continued in China and Vietnam and in 1942 Kubota was called on to survey the Lake Toba and Asahan River region of Northern Sumatra for the Japanese Occupation Forces.[11] It was this opportunity which first spurred in Kubota the concept of what was to become the Asahan project, involving an alumina refinery powered from a hydro-electricity scheme.

After the war, former members of the companies reassembled in Tokyo and Kubota formed the *Shinkō sangyō kensetsusha* in June 1946 with capital of 190,000 yen. Its name was changed to *Nippon kōei kabushiki kaisha* in October 1947. The company undertook several domestic reconstruction projects, including water resources, electricity and other civil engineering contracts. The company's postwar overseas operations began with a foreign tour in late 1953 by Kubota, which proved a successful gamble in opening up markets for the company and, it seems, resulted in a broader relationship between Japan and countries of Southeast Asia.

Kubota was not one to lose an opportunity. While in Burma on his 1953 trip, he was told by Burmese officials of United States consultants' reports on proposed electricity generation schemes and one, at Balu-Chaung, attracted his attention. Subsequently, on the plane to Europe he drafted, and from Paris sent, a letter requesting permission from the Burmese Government to survey the project further. The acceptance reached him in Mexico, whereupon he flew back to Burma in December 1953 and, as a result, a team of six Nippon Kōei engineers made an initial study, and a contract for the later stages of the project was signed in April 1954. The Burmese had originally intended to raise finance for the project themselves, but decided to request funds from the Japanese Government through a reparations agreement, and Kubota appears to have been largely responsible for persuading Japanese Prime Minister Yoshida and senior government officials of the value of using reparations to finance the Balu-Chaung development.[12] The project, costing 10,390 million yen,[13] was the largest single item in the reparations agreement and over a ten-year period Nippon Kōei not surprisingly was given consultancy and supervisory responsibility for all stages.

Kubota travelled to South Vietnam in 1955 and offered to survey the Da Nhim Dam scheme, part of the Mekong River development programme. This was a site he had seen during the war, when it was surveyed by the Japanese military administration. The Economic Commission for Asia and the Far East (ECAFE) gave Nippon Kōei the contract for the dam over a French company, but finance again proved a problem. The Export-Import Bank (Eximbank) was considered as a source of funds, but reparations talks between Japan and Vietnam were then in progress and it was decided to make Da Nhim the main reparations project. An Eximbank loan of 2,700 million yen was made in November 1960 for the purpose.[14] Between 1955 and 1964, Nippon Kōei saw the project through to its completion and followed this with a survey of the Da Nhim power station and repairs to it in 1971-2 under Japanese Government grant aid, and restoration of the Da Nhim-Saigon transmission line in 1973-5, also under grant aid. The early success of the company in South Vietnam assisted in Nippon Kōei being asked by ECAFE's Water Resources Bureau to undertake studies of a section of the Mekong Basin development scheme, which led eventually to active Japanese participation in the Mekong Committee.[15]

Work done for the United Nations helped extend Nippon Kōei's links with Laos initiated at the 1955 Tokyo meeting of the ECAFE. The company was selected to undertake feasibility studies in the Upper and Lower Nam Ngum River, and these were made between 1959 and 1962 through a contribution from the United Nations Development Programme (UNDP). After the creation of the Nam Ngum Development Fund in 1966, Nippon Kōei was retained as engineering consultant and carried out surveys, design and construction supervision of the Nam Ngum hydro-electric project. The Japanese Government made a grant to Laos for the fund in 1966 and loans in 1974 and 1976.

This Laos connection was maintained. Japanese grants for extensions to Vientiane airport in 1969 and 1970 were the results of a Nippon Kōei feasibility study and were implemented under its supervision. Other projects were carried out for the Mekong Committee and for the Asian Development Bank (ADB), and a Japanese Government grant for refugee resettlement in Na Phok was contracted to Nippon Kōei in 1973 for design and construction supervision.

Links established with the UNDP through Laotian projects helped Nippon Kōei gain the contract for survey of the Karnali Dam site in Nepal from 1962 to 1965. The company carried out a pre-feasibility study on the Kulikhani hydro-electric scheme under commission from the OTCA in 1962-3 and completed the project with a further study in 1973-4 sponsored by the Japanese Government, and design and supervision under a loan from the International Development Association (IDA) in 1975. Nippon Kōei also assisted in a study of Janakpur district agriculture sponsored by the OTCA, which led to a grant of machinery from the Japanese Government worth 45 million yen in 1972.

Nippon Kōei's ties with Indonesia — the strongest with any of its country clients — were initiated by Kubota's wartime experience in the country, especially his enthusiasm for developing the Asahan area, and consolidated by the reparations agreement with Indonesia concluded in 1958. In fact, as Nishihara shows, 'Kubota's company . . . gained nearly an exclusive hold over Indonesia's infrastructure projects under the reparations fund.'[16] It did this together with Kajima Construction Company, which built the projects. Kubota's desire to develop the Asahan region in Sumatra was not fulfilled under reparations, but his work on the so-called '3K dams' and development of the Brantas River region in East Java laid the foundation for a long and profitable relationship between the company and the Indonesian Government, and a widening aid relationship between Japan and Indonesia.

Nishihara argues that major projects under the reparations agreements were first proposed by 'private experts' outside the official reparations negotiations. He claims that Kubota lobbied successfully with high Indonesian officials, including President Sukarno and others in the Ministry of Public Works, to gain the contracts.[17] The first reparations projects involved surveys, design and supervision of the construction of the South Tulungagung reclamation scheme near Surabaya in East Java, which included construction of a new tunnel for the diversion of water from the Brantas River. An earlier tunnel built by the Japanese during the war had proved inefficient and Nippon

Kōei carried out the new work between 1959 and 1961. It became the first of a number of other projects in the Brantas region.

Two of the 3K dams — those on the Karangkates and Konto Rivers, both tributaries of the Brantas — were connected to this development. The third, the Riam Kanan Dam, was located in South Kalimantan. All three were designed for hydro-electric generation. Studies were begun in 1959 for the Karangkates project and in 1961 for the Kali Konto and Riam Kanan Dams. Work on them was not completed until 1973, however, despite expectations of 1967 as a target date, and the high cost necessitated further IGGI-base loans in 1968, 1969 and 1973 totalling 16,398 million yen. This huge expenditure and the delays in construction led to some criticism of these particular projects and of the methods employed by Nippon Kōei in supervising the work.[18]

Despite delays, other projects in the East Java region fell to Nippon Kōei tender, a connection which has continued to the late 1970s. The company undertook survey and design in 1961-3 for the Wlingi Dam project, situated close to the Karangkates Dam. Japanese Government loans were extended in 1975 and 1976 to enable it to be completed and Nippon Kōei again supervised construction. The firm surveyed and then completed the Kali Porong River improvement project (loans 1970 and 1976) and Kali Surabaya improvement (loans 1974 and 1976) and undertook a survey in 1971-2 commissioned by the OTCA of the Brantas River Basin. Other water resources and power projects were also completed: Riam Kanan power transmission (loan 1972), Wonogiri Dam in central Java (loan 1975), Bengawan Solo River Basin project (OTCA study 1972) and Way Umpu and Way Pengubuan (South Sumatra) irrigation (loan 1974).

The same intense pattern of Nippon Kōei operations was seen in South Korea, where Kubota and his staff had had successes in dam construction before the war. The Chunchon and Sumjinkang hyro-electric projects were completed in 1962 and 1965 and in 1962 they undertook a survey for the So Yang Gang Dam, which was taken up as a reparations project in 1965 and included in the Second Year Economic Cooperation Plan. Japanese loans were extended first in August 1967 and later in 1968 and 1970. Nippon Kōei won the contract for the dam. It also tendered successfully for the Taechung multipurpose dam project, for which a loan was made in 1974.

Nippon Kōei was one of the first Japanese consultant groups to go into Africa after the war, and its work there in the mid-1960s repre-sented the first efforts of the newly established ECFA. Kubota's personal connections helped there as they had done in other countries.

Nagatsuka claims that President Sukarno offered in 1963 to introduce Kubota to President Nkrumah[19] and the company history suggests the same. Assistance was gained from Ambassador Oki Munetoshi and from the Japanese Embassy in Accra, and an invitation from the President arrived for Kubota through diplomatic channels.[20] A Nippon Kōei-ECFA team went to Ghana in 1964 and completed preliminary studies of the White Volta River. Kubota flew to meet Nkrumah in Cairo and a letter of intent from the Ghanaian Government followed soon after. Arrangements were also made for the Japanese Government through the Overseas Economic Cooperation Fund (OECF) to extend long-term credit for further surveys, the first time this kind of concessional finance had been made available for consultancy work.

Consultants and Foreign Aid

In 1978 there were nearly 50 member firms of the ECFA, the main association of engineering consultants in Japan. They embraced all specialities, from general civil engineering to mapping, urban planning and concrete engineering.[21] Over half of the companies employed between 51 and 200 specialist staff and capitalisation was generally low, over 30 companies having assets of less than 100 million yen. In 1972, only three companies possessed capital of over 1,000 million yen. In international terms, Japanese consultants were said to be weak competitors, gaining only 2.9 per cent of UNDP contracts between 1959 and 1970, 7.3 per cent of ADB contracts up to September 1971 and only 0.6 per cent of World Bank Group contracts between 1966 and 1970. This 'weakness' theme has been reiterated in several official reports, including successive MITI economic cooperation annuals, the 1978 MFA report and recommendations from the Advisory Council on Overseas Economic Cooperation. In fact, about half of the companies relied on overseas operations for less than 10 per cent of their business, while others depended for over half their work on Japanese Government contracts.

The reasons for the diversity of consultancy firms lay in the problems they faced as an industry and as individual companies. There was no law regulating the industry, although companies were eligible for certain taxation concessions and export insurance. The allegedly poor international sense of the Japanese was also cited as working against the recruitment of professional staff willing to work overseas, although this may have had more to do with differences in company structure and tradition — a preference for home-based careers — than ethnocentrism. The poor foreign linguistic capabilities of some companies could be a damper on their overseas business and the Japanese university system was said to train economists and engineers too specialised for general consulting.[22] The low capital base of companies lessened opportunities for large-scale work overseas and the gradual untying of Japanese aid from donor country procurement made their position more uncertain.

The ECFA was one of eleven or more consulting firm associations in Japan and was one of the most vigorous lobbyists for the consultants' cause. The other large group, the Japan Consulting Institute, was set up in 1957 to promote exports of heavy engineering equipment. It was absorbed into the Japan External Trade Organisation (JETRO) in April 1971 and until then had been very successful in securing contracts for exports of Japanese plant and equipment. The ECFA was the main spokesman for consultants in the late 1970s and through it was channelled the bulk of government financial assistance to consultants.

The ECFA called itself 'an information centre linking clients abroad with ECFA member consulting firms'. It provided also an additional source of advice to government ministries and agencies considering aid project implementation. It made surveys overseas at its own expense (but with MITI subsidy) and encouraged member firms to initiate projects themselves. The ECFA was in fact set up on the understanding that government ministries would support its financing. The then Deputy Director of MITI's Technical Cooperation Division, Yamaguchi Jinshū, was one of a small group in and around MITI who urged taxation concessions and government funding of the proposed group; these measures were implemented from fiscal 1964. Yamaguchi became the ECFA's Director in 1964, a position which he has occupied since.

The ECFA was aggressive both in identifying projects and in assisting members to secure contracts. In its own words, 'to support member companies in winning contracts for development projects, we send survey teams to various countries and conduct site surveys, gather information and *undertake preliminary negotiations*'.[23] ECFA members, according to the association's report, lobbied developing-country governments with proposals on projects identified by themselves. Survey teams were composed of members either from one company alone or from several member firms. Project identification was especially designed to help members bring new projects to the attention of developing-country governments, in the hope of tendering successfully for the contract at a later date. Of the 984 surveys undertaken between

1964 and 1974, 100 or about 10 per cent — according to ECFA, a figure high enough to indicate 'success' over the years — led to a member firm securing the contract.

Of 1,165 projects listed in the association's *History* as surveyed between 1964 and 1973, there was a clear bias towards Asia. 644 were in that region, followed by 223 in the Middle East, 127 in the Americas, 110 in Africa, 21 in Europe, 21 in Oceania and 19 for international agencies. Within this distribution, Indonesia was the country with the largest number of projects assessed (127), followed by the Philippines (115) and Iran (78). In all, teams were sent to 84 different countries, although it seems that ECFA assistance benefited a few companies in particular among the 47 members. Over the period, with the support of the ECFA, Nippon Kōei took part in 164 project surveys (14.1 per cent), Pacific Consultants International (PCI) 207 (17.8 per cent) and Sanyū Consultants 273 (23.4 per cent).[24]

Success in tendering followed in ratios closely proportionate to company survey efforts. Of 81 major contracts gained by members between 1964 and 1973, 22 were won by Nippon Kōei (27.2 per cent), 13 by PCI (16 per cent) and 20 by Sanyū (24.7 per cent). At the same time, Sanyū, for example, concentrated its effort in only five countries: the Philippines, Indonesia, East Pakistan (Bangladesh), Afghanistan and Iran. PCI was heavily involved in the Middle East, while Nippon Kōei spread its resources across many countries, which certainly proved profitable.

For consultants, aid projects were not an end in themselves but simply a means to increase profit and further business opportunity. However, life as an overseas consultant (as a number of executives put it) was not easy and earnings were hard won. Survival demanded a constant generation of new development projects and possibilities, which was why a company like Sanyū sent out so many project identification teams and why the investment of men, money and time in developing familiarity with particular countries or regions and their governments was necessary.

Overseas consulting was an industry created by foreign aid. Not only did the Japanese Government support it with contracts, but it also gave the industry direct assistance, which was itself classified as a component of the aid programme (as technical aid); consultants therefore contributed to Japan's aid performance in two ways. The IDC depended on work from JICA to operate,[25] and the ECFA relied for about one-third of its budget on contributions from the MITI technical aid programme. This assistance rose steadily from 15 million yen in 1964 to 116 million yen in 1976. Other consulting associations (such as the Overseas Transport Consultants Association and the Overseas Construction Association) also received government subsidies to assist their work.

Financial support was only one facet of government policy to promote the consulting industry. The ECFA always ensured that the interests of the industry were well publicised; the diary of ECFA activities between April 1964 and December 1973[26] recorded constant meetings between association officials and bureaucrats in all related ministries; the government could not possibly have been unaware of the ECFA's presence or purpose. In fact, it readily acknowledged the ECFA's communications function in identifying and preparing projects.

Government assistance also allowed consultants to participate in official JICA surveys, inclusion of consultants under provisions for export insurance, taxation concessions (created in a law passed for the purpose in April 1964, soon after the ECFA was established), debt insurance and, more recently, bond insurance. There seemed, however, to be few concrete proposals beyond these. The 1977 MITI economic cooperation report referred to the need for assistance to the industry in the light of increasingly risky consulting tasks,[27] while the Advisory Council on Overseas Economic Cooperation in 1975 suggested rather tamely that 'in future, it is desirable that consultants obtain work from the first surveys and planning through to design and construction supervision. This can improve consistency in projects'.[28] The 1978 MFA report on development assistance pointed to the increasing scale and cost of surveys as one of the major problems in Japan's technical assistance, although the MFA remained firmly wedded to a policy of promoting the industry.[29] The recommendations on the consulting industry made in the 1971 report on technical cooperation evoked no new initiatives from the ministries. The need seen then[30] for inexperienced companies to gain more overseas work by a sharing of JICA commissions was not satisfied, as we shall see.

The constructive attempts in 1976 by the ECFA and by other consultant associations (later partly acceded to) to persuade the Japanese Government to give more direct loans for consultancies ('engineering loans') not only publicised their own cause but also touched some raw official nerves because of rivalries between the OECF and JICA regarding the implementation of surveys.[31] The participation of consultants in debate about the aid administration revealed how close they were to aid policy. While aid was for them a means to make money, they had become essential to the aid process: they grew on aid, aid grew because of them, and their energy helped compensate for the conservatism and inertia of the Japanese aid bureaucracy.

Arranging Surveys

Development surveys, as we said, were carried out mainly under technical assistance programmes administered by JICA. The later stages of the project — detailed design and construction supervision — were completed under loan agreements, the recipient government assuming formal control of the letting and management of contracts. These two separate aspects of project implementation were not, however, unrelated and, as seen in connection with Nippon Kōei, consulting work was often the link in all stages of projects as well as across projects. Causality, however, is elusive: was there a direct relationship between consultants and the development of the aid process, or were consultants its servants?

While the phases of surveying a project were in theory straightforward — general pre-feasibility assessment, feasibility study, cost-benefit analysis, design and construction — one senior JICA official, Tanaka Tsuneo, pointed out that in practice it was difficult for this order to be retained. There was a tendency, especially with projects becoming larger, for the project's feasibility to be assessed before complete surveys had been finalised, for funds to be committed before surveys had been made, or for surveys to be conducted according to cost-benefit rankings of projects irrespective of the social and economic implications. Tanaka admitted quite frankly that these inconsistencies occurred for several reasons, including the political nature of recipient requests, the eagerness of private interests for profit and the lack of government policy on survey methods, such as the World Bank and others had instituted (although the government was actively developing a set of guidelines for survey procedures).[32]

Political and economic constraints on surveys were serious indeed and the way these operated was directly relevant to bilateral aid policies. They will be considered in Chapter 8. There were, however, structural problems which posed barriers to what JICA saw as appropriate reforms. The 1974 annual JICA report cited a need for survey budgets to be increased before qualitative improvement was possible, more planning of aid (including country programming, better use of international resources etc.), proper use of consultants by 'proposal contracting' and greater consistency in official supervision. A more fundamental issue identified was government decision-making itself, for efficient surveying and project assessment depended on the rational and scientific use of information, in conjunction with country planning accepted by all sections of the administration.

The choice of consultant for feasibility studies frequently determined the progress of projects and the stage of bilateral aid relationships. For surveys done under technical cooperation programmes (project finding, pre-feasibility, feasibility), JICA was the managing agency, although the Mining and Metals Agency also carried out some in mining and prospecting. Consultants were chosen to participate in project surveys on fairly flexible criteria: there was no actual tender system in operation, although both officials and the industry recognised that tendering was a possible alternative. Some of the larger and more experienced firms, being accustomed to tendering, did not oppose it. Surveys involving private consultants were supervised by a committee, composed of officials from the competent ministry and from relevant JICA departments. JICA provided the administrative backup for this committee.

In selecting the consultant to carry out a survey, it was not the committees which were responsible, but JICA. The choice, however, was the result of discussion between committee members and JICA personnel and, of course, prospective consultants. The committee's effectiveness could differ according to the area of work. Agricultural and Mining and Industry Departments maintained supervisory committees only for 'important' projects, that is large and expensive undertakings (such as the Purari River surveys in Papua New Guinea) or those linked to government loans or having diplomatic considerations. Sources, however, from the Social Development Cooperation Department put a figure of 90 per cent on consultant surveys with supervisory committees.[33]

Feasibility studies were commissioned by 'invitation proposals', for which JICA approached a consultant group. Invitations were made after pre-feasibility studies done by JICA teams had been finalised and JICA's own analysis completed. JICA's public attitude was one of 'fairness' consistent with the policy of the Japanese Government of promoting the development of the whole consultant industry. Fairness, however, was tempered by an appreciation of the type of survey involved and of where it was to be carried out.

Analysis of surveys listed by JICA as completed by itself and by its predecessor, the OTCA, between 1962 and 1973, showed a reasonable spread of work between the main consultants in different sectors. While the largest general consultants carried out infrastructure project surveys (electrification, irrigation, water resources, roads, bridges, etc.), surveys for specialised projects were contracted to firms catering to that restricted demand. Thus Japan Airport Consultants Incorporated completed surveys for airport projects (such as in Vientiane), Mitsui Kinzoku Engineering and others were used for mining and prospecting studies, Universal Marine Consultants undertook marine resources surveys, Pacific Aero or Asia Air Survey did mapping, while Nihon Suidō Consultants or Tokyo Engineering Consultants were commissioned to study urban water and sewerage systems.

Government promotion of consultants was intended to build up the experience of the industry by encouraging the development of less experienced firms, but attitudes about the low quality of consultants were entrenched in the bureaucracy. MFA officials complained that consultants needed constant supervision and that the poor work produced confirmed that JICA was the most appropriate surveying agency. The MAFF regarded only five or six companies as being effective in agricultural projects, while MITI officials admitted that a wide performance gap existed between large and small firms in spite of government policies to narrow it. A company's experience, therefore, remained significant in the choice of consultant, notwithstanding the official attitude against resorting to that criterion.

Experience, however, was a broad notion and could be judged in many ways. Experience in the kind of project and in the country in question was recognised as essential. This related to the company's previous jobs and to its familiarity with the country, its language and even with the district in which the project was planned. The ability of the company to project itself as specialising in certain fields or countries was often decisive. Nippon Kōei, as we have seen, built its business on proven skills in water resources engineering and on an association with Indonesia, South Korea, Vietnam and Nepal. This helped it accumulate local knowledge, a reputation within the recipient government and a foreign linguistic capacity within the firm. The establishment of branch offices in the country also represented a longer-term commitment by the company to its work there, and an institutionalisation of its local presence.

As the example of Nippon Kōei showed, country experience was cumulative, leading to a steady stream of projects and contracts by companies over a period of years. Project finding was crucial in company development and consultant firms set up as subsidiaries of larger industrial or trading concerns (about half of the total) were able to draw on contracts from parent and related companies. Apart from these obvious ties, such firms relied on companies within their group for information about likely jobs. Consultants associated with trading companies were particularly assisted in this respect.

Independent firms faced a more difficult task in building overseas expertise. One company, founded in 1962 and in 1976 one of the most respected firms of engineering consultants in Japan, specialised in desert irrigation and was recognised as being best equipped to handle civil engineering projects in Iran. This reputation, however, took over ten years to build, and the company admitted to having fostered relations of trust and confidence with the Iranian Government 'by the expenditure of large amounts of money and time'. Past effort spent in cultivating younger bureaucrats, especially in the Water and Power Ministry, and on-the-job demonstration of the company's own expertise, paid off in easy relations with important ministers in the Shah's government. This company claimed to have 20 to 30 engineers on its staff with experience in and knowledge of Iranian conditions.[34]

This firm's second overseas survey in 1965, that for the Taleghan irrigation scheme, was funded by the OTCA and was the agency's first survey in Iran. The company carried that project through to its construction stage. It was also involved in OTCA/JICA's agricultural irrigation project at Sistan from 1968. The firm claimed success in several South Korean projects (notably the Yong San Gang irrigation development, the initial survey for which was requested by South Korea's Agricultural Development Corporation) and a contract to survey the construction of Cairo's water supply. A request for yen loans to carry this out was approved by the Japanese Government in 1976.

Choosing a consultant for a government-sponsored survey thus required detailed investigation of the background of the country's request to the Japanese Government, for relationships between projects and companies were often crucial in the decision. For that reason, the Director of JICA's Survey and Planning Department, Tanaka Tsuneo, cited the need to gain detailed information on proposals as the most important requirement for choosing a consultant. He saw this to be necessary because of the various political and economic incentives for requesting: the recipient's own political and economic judgements; assessments in Japan of the need to rectify trade imbalances; diplomatic necessity; opinions of visiting technical advisers; suggestions by international organisations; and profit-seeking by private enterprise.[35]

The information Tanaka required, but often gathered in insufficient depth by the government, included the role of the proposed survey in the LDC's national plans, details of the survey itself, cost estimates, counterpart administration and so forth. Specifically, officials needed to know how the project to be surveyed was first identified and by whom, for although consultant firms were valued for their professional neutrality and objectivity, strict maintenance of those ethical standards was not always possible in a competitive industry. In fact, the corporate links between 20 or so consultants and other private enterprises suggested preferences for close association between consultants and affiliated companies with overseas interests. Many relied on affiliates to generate the bulk of their work, at least in the early years. Parent firms might also prefer consultants close to them to complete surveys, thus assisting in the parent's successful tendering if a project were found to be feasible.[36]

Information, therefore, was essential to both the consultants and the government. A consultant's intelligence was best gained at the earliest stages of project identification, and missions sponsored by the ECFA were useful in this regard. In the final choice of a company to carry out a survey, those which knew the project had a distinct advantage. This was accepted by officials in all JICA departments associated with survey work, by ministry officials and by consultants themselves.

After knowledge of country, even locality, and experience, it was advantageous to be connected with the project in some way. Application of all of these criteria applied especially to surveys in civil engineering fields, where competition between large firms of consultants was tough. Bigger firms found it far easier to accumulate inside knowledge. They had the resources to tap intelligence in many countries and had lines of communication to both recipient and Japanese governments. The lack of sophisticated data-handling channels in the Japanese aid administration left smaller consultants at a definite disadvantage in the competition for contracts. Companies which had been associated with a project since its inception were hard to beat.

The choice was therefore dependent on factors other than objective assessment of the relative skills of likely contractors. Consultants realised the benefits of project finding, although smaller firms found the investment in that kind of exercise prohibitive, despite ECFA assistance. One of the largest companies undertook up to six special project finding surveys each year and many more project identification surveys. The small capital base of consultants (relative to other industries) demanded their constant efforts at identifying and promoting projects. Officials admitted that most projects surveyed had their origins in the initiatives of private enterprise, although a precise evaluation of that claim was difficult.

When a company wanted the Japanese Government to take up a project requested by the recipient country, the firm made representations to JICA and to the ministries. The consultant for a survey was chosen by the relevant JICA department after discussions with the associated ministry. Which ministry depended on the type of survey envisaged, for the MAFF vetted agricultural surveys, MITI those for mining and industry, Construction those for public works and so on.[37] While surveys to be undertaken in any one year were decided tentatively before the budget request for that year, changes were made after budget allocation and decisions about which surveys would be carried out and by whom were not made for budgeting. On the other hand, as the representatives of one firm of consultants put it, commonsense business practice and their own forward accounting demanded that ministries were made aware of company interests. It became fairly evident which projects companies would expect to receive in the following financial year.

The specialist ministries had their own preferential lists of consultants, 'their own standards and favourite companies', and it seems that the ministry's choice was a telling factor in the final decision. JICA officials claimed that they in fact had the formal responsibility in choosing and thereby could refuse work to consultants close to particular ministries, in order to uphold the government policy of fairness in letting contracts. They admitted, however, that such refusals were few; the strength of ministry opinions within JICA departments lessened the chances of conflict considerably.

Exchange of information was greatly assisted by the ECFA, especially in its sponsorship of project identification missions, but it was not altogether clear what role the ECFA, and other consultant associations, exercised in decisions about projects. Certainly the associations participated in the daily round of informal discussions on projects and surveys, and the MFA and other ministries were careful to keep informed of the progress of studies and projects sponsored by the ECFA. This was part of the ministries' aid intelligence network. Officials of specialist ministries, especially MITI and Construction,[38] were the most frequent points of contact, and MITI had an officer assigned specifically to ECFA liaison. Ministries were careful to point out, however, that the ECFA was not part of the formal decision-making process.

The MAFF, in contrast, did seek the recommendation on consultants of the Agricultural Development Consultants' Association (*Nōgyō kaihatsu konsarutantsu kyōkai*) after negotiations with companies through the association.[39] The advice was passed along to JICA to assist in the final decision, although the MAFF, as we said earlier, considered that any choice would be between only a few companies, on the basis of experience in and relation to the project involved.

On balance, the ECFA and its sister associations were important in the early stages of projects and, despite their primary functions as coordinating and promotional bodies, they could influence the climate of opinion about appropriate consultants for proposed surveys. This extended to discussions with responsible officers in JICA and the ministries.

Completing Projects

The later stages of projects again provided work for engineering consultants, but were usually funded by government loans or grants rather than technical assistance. It is relevant here to consider whether later stages of projects were in fact carried out by the same group which undertook feasibility studies or even project identification studies.

It was in a consultant firm's interest to manage all stages of a project, but it is not certain how often this happened and it was not possible, on the available evidence, to calculate how many loan projects were managed in this way. Since loan contracts were put out to international tender, it was not always easy for Japanese tenders to win. Recipient government policy was important: some preferred later stages of projects to be financed by the same donor which provided the initial surveys. Some also preferred the same consultants to advise, even though official policy made no such provision. According to JICA documents, Japanese grant aid procedure specifically required that the same consulting firm which undertook preliminary design would carry out design of construction (partly because of the possibility of delays and extra cost if another consultant were to take over).

Some Japanese officials declared that few projects were handled by one company through all stages, while others considered it to be quite common. Examination of the project listings of selected consultants over several years suggests that while the 'bread and butter' of all consultants' work was the standard feasibility study, the older and bigger firms with overall skills (Nippon Kōei, Sanyū, PIC, Electric Power Development Company, Nippon Telecommunications) won contracts for later stages, as did specialists like Universal Marine.[40]

It is likely that the biases already existing in the system in favour of a few large companies — due to experience, resources available for the development of project ideas, and Japanese officials' attitudes to consultants — were strengthened at the later stages of projects. Despite donor-government supervision, it was not possible for policy on consultant promotion to be rigidly adhered to, since tenders were the responsibility of the recipient government. The principle of 'fairness' could be applied less easily. The success of consultants was dependent more on their own efforts at the recipient end, and in this respect the long-term associations of companies with recipients took on real significance. The connections between companies and projects were also of influence, although there was no guarantee that companies could successfully tender for a project in which they had invested time or money.

The Japanese Government was not isolated from this selection process. Tenders were called by the recipient government but the MFA and the OECF, in the case of Indonesia, for example, tried to ensure that at least one Japanese firm was included on any short list of tenderers. Likewise, in cases where capital grants were being extended, the MFA's Second Economic Cooperation Division and JICA assisted the recipient administration in contracting arrangements.

As business houses, consultants still worked behind the scenes to ensure participation in projects, given the officially cautious attitude to actively initiating aid. In one sense consultants acted as 'agents' for the Japanese Government in project finding and identification. These efforts helped both consultants and the OECF, for example, when there were understandings about the kinds of projects the government considered appropriate. It was up to consultants themselves to persuade governments of developing countries of the benefits of projects, and Japanese officials of their long-term worth. One consultant put it more colourfully as 'selling' ideas to recipients and 'coaxing' money from Japanese officials. Whatever the term, the bridging or communication function of consultants in early and later stages of projects was clear. Consultants claimed a spontaneity lacking in government, which depended on a government-to-government request as the start of aid work. Consultants were at the forefront of these relationships, pushing and probing for new business: commercial practice provided a momentum otherwise missing from the Japanese aid process.

Consultants were therefore a purposeful vehicle for the building of relations, at the business and government level, between developing countries and Japan. We noted how consultants often concentrated their efforts on building contacts with a few recipient governments. They were, of course, only one of a number of Japanese representatives, both official and private, in these nations, but were one of the more mobile and flexible groups. Although it meant that the focus of Japanese aid was narrowed onto the work of a few consultant companies operating in a handful of recipient countries, consultants were nevertheless instrumental in initiating aid to new recipients. Consultants, like their surveys themselves, had an intelligence function: in publicising projects they politicised the culling of requests but in so doing gave officials the information necessary on recipient thinking about projects and complemented the task of officials in ranking requests. The informal articulation of consultants' interests, in helping attach priorities to proposals, influenced the eventual allocation of aid funds, first of technical aid and, later as projects advanced to the construction stage, of loan funds. Consultants indeed often were an important link between the early and later stages of projects.

This gave coherence to aid policies (especially in connecting technical and capital aid), but at the same time led capital aid into predetermined paths. Allocation was frequently set by the manner in which a project had been identified and surveyed (and by whom), and by the energy of the company in pursuing its business in particular districts or countries. For that reason, the projects undertaken by consultants naturally revealed, or generated, new aid possibilities which companies were keen to promote. The consultants' aggressive search for such opportunities gave impetus to the aid process. The relationship, therefore, between aid and consultants was essentially one of interdependence. Consultants, as one of the most active parts of the whole aid system, took from the ministries much of the responsibility for initiating new policies, but did so because they were private enterprises, seeking to maximise profit in an unpredictable business where government policies were far from clear.

Notes

1. For an analysis of project administration see Dennis A. Rondinelli, 'International assistance policy and development project administration: the impact of imperious rationality', *International Organization*, vol. 30 (Autumn 1976), pp. 573-605.

2. See Kokusai kyōryoku jigyōdan, *Kokusai kyōryoku jigyōdan nenpō (Japan International Cooperation Agency yearbook)*, 1975 (hereafter *JICA 1975*), pp. 98-9.

3. Zaisei chōsakai, *Kuni no yosan (The nation's budget)* (Tokyo, Dōyū shobō, 1978).

4. Under Article 21.3 (e) of the JICA Law, the agency also undertook preliminary feasibility studies on projects when requested by private enterprises.

5. One MFA officer jokingly referred to the practice of ranking recipient countries according to their political importance to Japan, a type of aid policy 'Top Forty'.

6. Rondinelli, 'International assistance policy and development project administration', pp. 584-5.

7. A consultant engineer is a professional engineer who provides specialist knowledge and services to his client in return for a fee. He does not engage in actual production or construction work himself, although he may supervise others doing it. See Yamaguchi Jinshū, *Konsarutanto dokuhon* (*A consultant's reader*) (Tokyo, Kokusai kaihatsu jānarusha, 1976), Ch. 1.

8. As described in Nihon puranto kyōkai, *Nihon puranto kyōkai jūnenshi* (Japan Consulting Institute, *A ten-year history of the Japan Consulting Institute*) (Tokyo, May 1967).

9. See Yamaguchi, *Konsarutanto dokuhon*, pp. 149-50, and also his 'Hādo kara sofuto jidai o kangaeru' ('From the hard to the soft era: some thoughts'), *Kokusai kaihatsu jānaru*, 5 August 1972, pp. 4-8 at p. 5.

10. Material for this section was drawn from Kubota Yutaka and Yamaguchi Jinshū, *Ajia kaihatsu no kiban o kizuku: kaigai konsarutanto* (*Building the base for Asian development: overseas consultants*) (Tokyo, Ajia keizai kenkyūjo, 1967); Nagatsuka Riichi, *Kubota Yutaka 1966* (Tokyo, Denki jōhōsha, 1966); and Daiyamondosha, *Nippon kōei: kunizukuri no kishu to shite* (*Nippon Kōei: the standard bearer for nation building*) (Tokyo, Daiyamondosha, 1971); the pamphlet by Nippon Kōei, *Nippon Kōei Company Limited: consulting engineers*, n.d., Engineering Consulting Firms Association, *Engineering Consulting Firms Association (ECFA) 1976-77* and Gaimushō keizai kyōryokukyoku, *Keizai kyōryoku kankei shiryō* (*Materials on economic cooperation*), July 1974. Kubota was born on 27 April 1890 and graduated from Tokyo Imperial University in civil engineering in 1914. He joined the Home Ministry but soon moved out to form his own company in 1921. In 1926 he joined the Korea Power Company, of which he later became President. He also became a director of the Japan Nitrogenous Company and the Yalu River Development Company. These connections lasted until the end of the war.

11. See Nishihara Masashi, *The Japanese and Sukarno's Indonesia: Tokyo-Jakarta Relations 1951-1966* (Honolulu, The University Press of Hawaii, 1976), especially pp. 65-7 on Kubota's role in early Japanese relations with Indonesia.

12. See Kubota and Yamaguchi, *Ajia kaihatsu no kiban o kizuku*, p. 46.

13. Gaimushō keizai kyōryokukyoku, *Biruma keizai kyōryoku chōsadan hōkokusho* (Ministry of Foreign Affairs, Economic Cooperation Bureau, *Report of the mission inquiring into economic cooperation with Burma*), July 1972, pp. 143-6.

14. Information on the early stages of the project is also contained in Kubota Yutaka, 'Da Nhim Hydroelectric Project – Current Developmental State and Future Prospect' (sic), *Keizai kyōryoku*, July 1961, pp. 5-7.

15. Japan's role in the Mekong Committee is appraised by Hasegawa Sukehiro in his *Japanese Foreign Aid: Policy and Practice* (New York, Praeger, 1975), pp. 103-5, and by Lawrence Olson in *Japan in Postwar Asia* (London, Pall Mall Press, 1970), pp. 218-23.

16. Nishihara, *The Japanese and Sukarno's Indonesia*, p. 103.

17. Nishihara, *The Japanese and Sukarno's Indonesia*, pp. 67 and 106.

18. For example, Ogawa Kunihiko, *Kuroi keizai kyōryoku: kono ajia no genjitsu o miyo* ('*Black' economic cooperation: the Asian situation*) (Tokyo, Shakai shimpō, 1974), pp. 103-8. He claims that the tender prices of machinery were artificially high, but provides no clear evidence.

19. Nagatsuka, *Kubota Yutaka 1966*, pp. 368 and 372-5.

20. Kubota and Yamaguchi, *Ajia kaihatsu no kiban o kizuku*, p. 63.

21. This paragraph is based on Yamaguchi, *Konsarutanto dokuhon*, pp. 154-64.

22. Interview with officers of a consulting firm, 7 December 1976.

23. Kaigai konsarutingu kigyō kyōkai, *ECFA 1975 gaiyō* (Engineering Consulting Firms Association, *ECFA 1975 outline*), hereafter *ECFA 1975 outline*, p. 4 (my emphasis).

24. Kaigai konsarutingu kigyō kyōkai, *Aru dantai no kiroku: kaigai konsarutingu kigyō kyōkai jūnenshi* (Engineering Consulting Firms Association, *One organisation's record: a ten-year history of the Engineering Consulting Firms Association*) (Tokyo, 1974), Appendix, pp. 63-124 is the source for the figures in this and the following paragraph. Hereafter it is referred to as *ECFA History*.

25. Interview with an executive of IDC, 1 September 1976.

26. See diary of events in *ECFA History*, Appendix, pp. 1-38.

27. Tsūshō sangyōshō, *Keizai kyōryoku no genjō to mondaiten* (Ministry of International Trade and Industry, *Economic cooperation: present situation and problems*) (Tokyo, Tsūshō sangyō chōsakai, 1977), pp. 180-1.

28. Taigai keizai kyōryoku shingikai, *Kongo no kaihatsu kyōryoku no suishin ni tsuite* (Advisory Council on Overseas Economic Cooperation, *On the promotion of future development cooperation*), 18 August 1975, p. 12.

29. Gaimushō keizai kyōryokukyokuchō-hen, *Keizai kyōryoku no genkyō to tembō: nambaku mondai to kaihatsu enjo* (Ministry of Foreign Affairs Economic Cooperation Bureau Director-General (ed.), *Economic cooperation: present situation and prospects: the North-South problem and development assistance*) (Tokyo, Kokusai kyōryoku suishin kyōkai, 1978), p. 771.

30. Naikaku sōri daijin kambō shingishitsu taigai keizai kyōryoku tantō jimushitsu, *Taigai keizai kyōryoku shingikai tōshin (kaihatsu tojōkoku ni taisuru gijutsu kyōryoku no kakujū kyōka no tame no shisaku no tsuite) no jisshi jōkyō nado ni kansuru shiryō* (Prime Minister's Office, Prime Minister's Secretariat, Councillors' Office, Overseas Economic Cooperation Desk, *Materials on the implementation status of the report of the Advisory Council on Overseas Economic Cooperation (Policies for strengthening technical cooperation to developing countries)*), 24 January 1975, pp. 56 ff.

31. *Nihon keizai shimbun*, 22 October 1976. The request recalled an argument between the OECF and JICA over which was better suited to undertake surveys. One complaint of JICA in 1976 was that the OECF took on further unnecessary surveys of a project already proved feasible. The OECF wanted to have more surveys done on a loans base to achieve greater coordination between surveys and development projects which it would eventually administer (interviews with a JICA official, 17 November, and an OECF official, 24 December 1976). Engineering loans were first made in 1978.

32. Tanaka Tsuneo, 'Kaihatsu chōsa to tekkaku na hantei' ('Development surveys and precise judgements'), *Kokusai kyōryoku*, no. 240 (September 1974), pp. 14-17 at p. 15.

33. Interview, 17 November 1976.

34. Interview with a senior company executive, 29 November 1976.

35. Tanaka Tsuneo, 'Kaihatsu chōsa no arikata ni tsuite' ('The correct techniques of development surveys'), *Kokusai kyōryoku*, no. 244 (January-February 1975), pp. 27-33, at p. 28.

36. Interview with an officer of a large trading company, 10 December 1976.

37. Jurisdiction, however, was not quite so simple: rubber projects were MITI's responsibility unless they were associated with plantations, where they became an MAFF charge. MITI took care of construction materials, although veneer was under MAFF jurisdiction, while building (or the use of such materials) was the responsibility of the Ministry of Construction.

38. The Ministry of Construction was engaged in extensive economic

cooperation activities: despatch of specialists, administration of training centres (such as in Thailand for road construction), project surveys, overseas student training in Japan, assistance to the construction industry in overseas work etc. For details, see Kensetsuchō keikakukyoku kokusai kyōryokushitsu, *Kokusai kyōryoku no genkyō* (Ministry of Construction, Planning Bureau, International Cooperation Office, *The present situation of international cooperation*), April 1976.

39. One former senior MAFF official, however, did not regard them as very useful (interview, 30 November 1976).

40. These are only indicative conclusions, deriving from information about 'major projects' undertaken by companies over various periods. Data was compiled from material provided by companies and the Engineering Consulting Firms Association, and Jōhō kikaku kenkyūjo, *76 nenpan keizai kyōryoku puranto yushutsu binran* (Information Planning Institute, *Economic Cooperation and plant exports handbook, 1976 edition*) (Tokyo, 1976), an extremely useful compendium of current projects and the names of successful tenderers.

7 MANAGING BILATERAL RELATIONS

Aid policy decisions were indeed a matter of 'pushing and coaxing'. Budgets and procedures were essential in determining the directions, quantities and terms of Japanese aid flows; likewise, consultants were a source of energy and initiative in aid policy, keeping particular projects and bilateral relationships well ahead in the queue for official approval. But business consultants were not the only innovative force in what was an extremely open (albeit dispersed) policy-making system; such was the multiplicity of leverage points, in fact, that management by officials was essential in giving some order to cross-cutting pressures. This chapter examines how bilateral pressures affected the direction of aid flows.

An Asian Policy?

The nature of influences on the aid process derived to a large extent from the traditions of Japanese aid, as well as from more tangible structural characteristics. Chapters 1 and 2 pointed out the importance of Asia in Japan's aid programme, particularly in the early years of export promotion and Japan's finding her feet as an aid donor. One writer has gone further, asserting that the concept of economic cooperation itself was defined by Japan's relations with Southeast Asia, that the need for markets in and raw materials from that region under-pinned Japanese assistance policy.[1] Of the relevance of this to the 1950s and 1960s there is no doubt, but Japanese aid today, as has been demonstrated, is directed increasingly outside the Asian region. In part, the growing size and complexity of the aid programme were responsible for this trend — the response to international calls to ease the plight of the most seriously affected nations (most of which lie outside South-east Asia), the gradual spread of commitments to social development programmes, resources and economic infrastructure projects outside Asia as Japan's total contribution grew — at the same time as shifts in foreign policy and resources diplomacy encouraged absolute and relative changes in Japan's aid distribution in the 1970s, towards the Middle East and Africa.

Yet, in aid, as in politics, *plus ça change plus c'est la même chose.*

Japan's informal and cultural roots in Asia are strong; some would say there is a natural affinity which spills easily over into economic and commercial intercourse.[2] There is serious debate about the strength of Japanese psychological identity with Asian peoples,[3] but it cannot be doubted that Japan's self-perception as a donor is that of an Asian nation. Japan was an Asian power before she was a world power. Certainly her Asian heritage is insular, set apart from mainland Asia and, on several occasions in recent history, adversarial. It can also be argued that as a foremost industrial power Japan stands apart from her Asian neighbours, or at least that Japan is still vacillating over her perceived global role.[4] Nevertheless, Japan's historical tendency to build her Asian future 'by alternately exploiting the ideals of Asianism and "escape from Asia" '[5] is reflected in policy even today. Her relations with the Association of Southeast Asian Nations provide ample evidence of a continuing commitment to Asia, for many of the same reasons — chiefly economic and security — which have described her whole postwar foreign policy. At the same time a certain emotionalism has characterised the espoused policies of both Mr Fukuda and Mr Ohira, the immediate past and present Prime Ministers, with their 'heart to heart' and 'people to people' policies.

What have been the implications of Japan's Asian perspective for aid policy? Apart from the most recent sincere promises to ASEAN, for example, on the continued commitment of Japan to that association's development, there was a more fundamental response — an acceptance of the 'international division of aid labour' concept. This notion — perhaps to some extent a rationalisation — was premised on the idea that Japan's best interests lay in assisting Asian development and that, by implication, aid ties with nations in Africa, the Middle East and Latin America, regions seen to be largely served by other donors, should be restrained.

Reality was, of course, somewhat removed from this passivist theory, but the theory has strong support. Official publications tend to propose regionalist definitions of Japan's national interests, where geographical proximity is the criterion for the strength of aid ties.[6] The late Minato Tetsurō put it even stronger when he wrote that 'Asia is Japan's electorate',[7] while a former director of the MFA's Economic Cooperation Bureau once admitted the 'while "division of labour" is perhaps not the best word, Japan is after all Asia's main support'.[8] While Japanese policy was subject to considerably more complex pressures — Japan's global power status, for example, was cited by the MFA as one reason for greater interest in economic development in areas outside Asia — even the 1978 MFA aid report spoke of how natural were Japan's close economic and political ties with ASEAN and was noncommittal on Japan's aid relations with other regions, saying only that there was 'greater concern' over this problem and that increased Asian cooperation was becoming 'more of an issue'. In fact, it added that those countries, being geographically distant from Japan, had 'little self-confidence in being able to absorb aid from Japan'.

The desire, furthermore, that aid should be visible (perhaps a natural donor response) was zealously expressed in Japanese policy. The kudos attributed to a donor was regarded by Japanese aid officials as a justification necessary to the aid process, and from this sprang the perception of aid as an exchange, or as 'cooperation'. There may have been cultural roots to this tendency, for Japanese social custom stressed *ninjō*, or compassion, an essential element of relationships between two people which developed together with bonds of obligation in a tight fabric of emotional accounting. This led to an overriding concern to foster a few close relationships at the expense of many others. There was a demand for compensation for favours and the avoidance of relations in which exchange was absent. One writer suggested that the weak Japanese understanding of *noblesse oblige* was a result of these social customs. That they were one influence on aid policy was confirmed by the stated relevance of *ninjō* to bilateral aid.[9]

Such attitudes affected aid relations with and limited Japanese understanding of Africa in particular. The African continent received, until the mid-1970s, only a small portion of Japanese official assistance (see Table 1.2). The first government loan was made to Uganda in 1966, followed by credits to Tanzania, Kenya and Nigeria in the same year. The country which received the greatest amount of Japanese loans was Egypt and, of Black African nations, Zaire. After Egypt, Nigeria was the largest recipient of Japanese ODA in Africa.[10] A sudden increase in the number of African loan recipients was not apparent until after 1973, coinciding with the visit of the Japanese Foreign Minister, Kimura Toshio, to Black Africa in late 1974 and with heightened Japanese diplomatic interest in the region.

Africa was regarded by aid officials in Japan as distant and difficult to deal with. They complained, rightly or wrongly, that Africans thought differently from Southeast Asians and that aid negotiations were correspondingly more protracted. They predicted that these difficulties would not soon diminish, despite growing aid flows to the region. Japanese knew little about Africa and about the conditions

Table 7.1: Geographical Distribution of Japanese Government Loans: Accumulated Value of Commitments up to Selected Years (Hundred Million Yen)

	1965		1968		1971		1974		1977 (September)	
	Value	%	Value	%	Value	%	Value	%	Value	%
East and South-east Asia	1287	40.5	3138	50.1	6654	58.7	12568	60.1	14598	56.4
South and West Asia	1584	49.7	2588	41.3	3784	33.4	5358	25.6	7260	28.0
Central and South America	253	7.9	301	4.8	531	4.7	724	3.5	921	3.6
Middle East and Europe*	61	1.9	79	1.3	209	1.8	1165	5.6	1193	4.6
Africa and Other	–	–	158	2.5	158	1.4	1099	5.2	1919	7.4
Total	3185	100.0	6264	100.0	11336	100.0	20915	100.0	25891	100.0

*Middle East and Europe includes Afghanistan, but not Morocco, Egypt, Algeria and the Sudan, which are grouped under 'Africa'. Yugoslavia is included because of loans in 1966 and 1972.

Source: To 1974, MITI, *Keizai kyōryoku no genjō to mondaiten*, 1975, Table 1-5, pp. 194-5; 1977, Gaimushō keizai kyōryokukyokuchō-hen, *Namboku mondai to kaihatsu enjo*, pp. 478-81.

Table 7.2: The Ten Main Recipients of Japanese Government Loans: Accumulated Value of Commitments up to Selected Years (Hundred Million Yen)

1965		1968		1971		1974		1977 (September)	
India	1188	India	1537	India	2346	Indonesia	5175	Indonesia	6579
South Korea	720	South Korea	1197	South Korea	2312	India	3227	India	4204
Taiwan	540	Pakistan	810	Indonesia	2255	South Korea	3134	South Korea	2627
Pakistan	378	Indonesia	747	Pakistan	1111	Pakistan	1449	Pakistan	1582
Brazil	180	Taiwan	540	Taiwan	621	Malaysia	900	Thailand	1273
Iran	61	Brazil	228	Malaysia	540	Philippines	866	Philippines	1147
Argentine	37	Thailand	216	Philippines	347	Thailand	856	Bangladesh	1074
South Vietnam	27	Malaysia	180	Peru	230	Iraq	745	Burma	953
Chile	22	Philippines }	108	Brazil	228	Taiwan	621	Malaysia	900
Sri Lanka	18	Burma } Nigeria }		Burma	190	Burma	538	Egypt	752

Source: As for Table 7.1.

upon which aid requests were made, and loan officials cited this as one reason why decisions on aid to Africa might take much longer than on similar requests from Asian countries. At a more materialistic level, they perceived that trading benefits to Japan from aid to Africa were insufficient to warrant a shift in priorities. Africa was regarded as something like the 'dark continent' into which Japanese aid disappeared with no acknowledgement of its origins. Some multilateral aid officials saw the African Development Fund as offering Japanese multilateral assistance far less visibility than did, for example, the Asian Development Bank. In short, prevailing attitudes towards countries in Africa as recipients only strengthened the bias in favour of Asia and the established decision-making and information-gathering procedures. Patterns of aid proved hard to redirect.

Not only was Japan's aid policy Asian-centred, but several related characteristics of aid policy-making encouraged a preoccupation with loan aid: first, the donor-oriented emphasis on 'economic cooperation' rather than on 'aid', the easiest way of constructing a programme identified with the view of no single ministry, led to a preference for government loans. This was further fostered by MITI's stress in aid management on the linkage between trade and aid, the desire by the MOF to minimise the economic burden on Japan's finances, and by bilateral concerns pursued by the MFA's regional bureaus in assessing aid requests. Again, career structures hindered the growth of special expertise and led to a tendency to focus on the details of the aid programme, on individual projects and cases, rather than on the general picture. This micro-approach resulted from the absence of policy guide-lines, the importance of budgeting in all areas of policy and the diversity of unrelated procedures for deciding on and managing aid. Finally, the low political relevance of aid and, correspondingly, the switching of political interest to types of aid useful to contractors and pressure groups meant a strong preference for project aid.

The effect of these biases showed clearly in the functional distribution of Japanese aid. Bilateral loans formed 33.1 per cent of total Japanese ODA in 1960, rose to 54.7 per cent in 1970 and dropped slightly to 51.8 per cent in 1978. Between 1970 and 1978, loans fell below 50 per cent of ODA only in 1977 (see Table 1.1). According to Table 1.2, furthermore, official aid flowed in the main to Asia, especially to Southeast Asia. The range of recipients of government loans expanded only slowly after 1957, when loans were first made. In the late 1960s new Asian recipients increased but it was only after 1973 that the number of recipients in regions other than Asia grew rapidly and, by 1977, 15 countries in Africa had received direct loans, rivalling the 17 in Asia. In value (Table 7.1), however, the African share (7.4 per cent) paled beside that of Asian countries (84.4 per cent).

A startling 85 per cent of all loans made by the Overseas Economic Cooperation Fund (OECF) until the end of 1975 went to Asia, 78.1 per cent going to six countries alone (Indonesia, South Korea, the Philippines, Thailand, Burma and Malaysia). According to OECF documents, exactly the same trend continued over the period 1976-8, although after Asia Africa became the next highest recipient region (12 per cent). The ten main recipients of Japanese (OECF and Export-Import Bank) loans made between 1957 and 1977 were, in order, Indonesia, India, South Korea, Pakistan, Thailand, the Philippines, Bangladesh, Burma, Malaysia and Egypt, all but one being Asian nations (see Table 7.2). Indeed, Tables 7.1 and 7.2 show what little change there was between 1965 and 1974 in the pattern of recipient regions and nations. The accumulated value of yen loans as distributed geographically was concentrated in South Asia up to 1960 but by 1965 the Southeast Asian tally nearly matched this. After 1965, Southeast Asia was well ahead as the leading recipient region and the Middle East and Africa only became prominent in the first half of the 1970s. The top recipient nations have seldom varied, with India, South Korea, Pakistan and, after 1965, Indonesia always being the leading four. Minor placings show Taiwan, Malaysia, the Philippines, Thailand and Burma consolidating their positions over the latter half of the 1960s and into the 1970s, while others, such as Nigeria, Brazil, Iraq and Chile, entered the list with occasional loans. Those countries whose place as important recipients resulted from the gradual accumulation of loan commitments from Japan were all Asian, and Indonesia's rise was perhaps the most remarkable.[11]

The strength of Japan's ties with a few countries was confirmed by her ranking as a donor in their total aid recipts. In 1975, according to MITI, Japan was the largest contributor of aid to Indonesia, Burma, Thailand, Malaysia, the Philippines and Peru and ranked first with the United States in aiding South Korea. Japan's position as a supplier of aid was high in Asian countries but low in Africa, where French and British aid predominated. In some South American countries (such as
⋯⋯ erformance ⋯ ⋯ved in the mid-1970s. In fact,
⋯⋯directed in greatest quantities
⋯⋯e of their foreign aid
⋯⋯aid relationships.
⋯⋯ans up until

recent years suggests an inflexibility of bilateral flows and a preference on the part of Japanese officials for a reactive, rather than an active, donor role.[12] The demand for government-to-government requests to initiate aid was common to many donors, but indicative also of caution and care to avoid being accused of intervening in LDC domestic affairs. Initiative in policy − even though more evident since 1975 − was inhibited by this preoccupation with existing aid forms, by a demand that aid given be visible, the lack of budget and country planning and by an acceptance of the 'international division of aid labour' concept.

Viewing Bilateral Relationships

David Wall concluded in his book, *The Charity of Nations*, that 'there are no objective criteria which can be used to determine the allocation of aid'. It is necessary, he added, for some people to 'be selected and called on to exercise their subjective judgement in determining how much (if any) should be made available, who should get it, in what forms, for what purposes and on what terms'.[13] These people, especially if they were bureaucrats, still had to operate within a fixed institutional environment. It was not simply a matter of balancing political, diplomatic or economic factors in favour of one policy course or decision, for influences were quite often erratic or operated in a way which confirmed identifiable objectives. The precise impact of aid on broad national policies or goals varied significantly across regions, countries, types of aid and methods of aid-giving.

International Effort and Aid: The IGGI

Japan was party to about 20 separate aid consortia, both active and inactive. These groups attempted to coordinate donor policies with the domestic economic policies of the recipient government, not necessarily determining future levels of aid from donors nor, as the Indian example revealed, always able to maintain the flow of aid at levels which the recipient desired, or even to assure donor concurrence with a recipient government's economic prio   14 Donor
dominate, for consortia
assessment of recipie
of international
from politi

Japan was a member of the Inter-Governmental Group on Indonesia (IGGI) from its inception in 1967 and always bore one of the largest donor burdens. 'The IGGI arrangement,' wrote Viviani, 'has been of crucial importance for the reconstruction of the Indonesian economy';[15] it was also essential to the development of Japanese-Indonesian aid relations and of Japan's aid policy itself. Through IGGI has been channelled the bulk of developed-country assistance for Indonesian economic development in the Suharto era. It was the most directly influential aid consortium of which Japan was a member, and Indonesia — Japan's largest single recipient — was the only one which regularly received a total volume commitment before the specific purposes of loans were agreed upon. The arrangement reflected a secure and self-satisfied approach to the bilateral relationship by donor and recipient alike and a structured bureaucratic response in Japan to the thorny problems of aid to a country both close and unknown. International and domestic pressures upon aid relations with Indonesia removed much management responsibility from officials, whose control of aid flows was weakened by the dependence of the IGGI system on adequate preparation of projects and requests for financing approvals.

Meetings of the IGGI were held annually in Amsterdam about the middle of each year. The Japanese team comprised officials of the MFA, the MOF and MITI and an OECF observer and was headed by senior officers of the MFA's Economic Cooperation Bureau. While there were bilateral talks preceding the IGGI conference until 1974, the only annual official contact after 1974 consisted of a visit by top Indonesian officials or ministers to Japan immediately before the Amsterdam conference. The former system involved Indonesian requests for total volume of aid, followed up by an OECF 'identification mission' whose report guided decisions on project selection.[16]

Under the present system projects were not selected until after the Japanese Government had fixed on its total commitment to the year's IGGI assistance and after this was pledged at the IGGI meeting. Projects were, however, never very far from the minds of member-government officials. The Indonesians favoured certain projects, as did Japanese ministries (even those not involved in loans decisions) and, as a government, Japan intended its pledge to be used to finance projects which it regarded as the most suitable (according to the criteria outlined in Chapter 4). Before IGGI convened, Japanese officials ranked candidate projects on the basis of two documents in particular: the World Bank report on Indonesian economic conditions, and the listing by Bappenas (the Indonesian planning authority) of 200 to 300 projects both being undertaken and yet to begin, although some Japanese officials complained that, of the latter, not many were 'good' projects, that is they did not meet Japanese project criteria. The Japanese Government fixed on a level of commitment shortly before the IGGI meeting, after consultations between the four ministries on the loans committee. According to officials, this decision was easy to reach, since the general parameters had already been established by the previous year's commitments, Indonesian demands, expected pledges from other donors and by the likelihood of Japan's supporting particular projects. The IGGI conference included discussions of Indonesia's economic outlook and projects by themselves were not an agenda item.

Official bilateral discussions between Indonesian and Japanese officials were held several months after the IGGI met. Before these talks, at which the projects for the year were settled, Japanese missions visited Indonesia to discuss possibilities and Indonesia, like any other recipient, sent an official request for projects through diplomatic channels to the Japanese Government. Requests from Jakarta reflected clear Indonesian Government policies on project development. As a general principle, according to Tokyo Embassy officials, Indonesia preferred the same donor which had financed feasibility studies on a project to take up any project loan. As was pointed out in Chapter 6, this policy helped strengthen bilateral ties and the links between technical aid for surveys and project loan aid. When projects were undertaken from start to finish by Japanese contractors the uncertainties in the bilateral relationship were lessened and aid flows became self-generating and self-reinforcing.

Bilateral meetings were devoted to the technical details and relative feasibility of projects, and at this conference-table venue the final shape of Japan's loan programme to Indonesia was determined. The options were greatly narrowed by this time, and although differences between Japanese and Indonesian representatives still appeared about how to finance which projects, the two governments agreed substantially on the functions of the IGGI and on the kinds of project to be funded through it.

Japan was well served by the IGGI, for most of her bilateral aid to Indonesia, indeed a large slice of her total government aid, was thus ostensibly removed from overt political pressure and benefited from the legitimacy it derived from an internationally coordinated aid programme. Criticism of non-IGGI Japanese aid to Indonesia, which had continued since the 1958-9 scandals surrounding reparations agreements, did not flow on to IGGI aid, except insofar as the total share of

Indonesia in Japan's development assistance was seen by some, even by Japanese officials (notably in the MOF), to be too large. Standard Japanese Government criteria for loans projects were not abrogated in choosing suitable projects for Japan's IGGI aid (nor was the usual pre-request promotion of projects), which indicated that the objectives of donor and recipient overlapped considerably. Without doubt, the types of projects financed by Japan through IGGI demonstrated the usefulness of the group, if defined (as by Posthumus) as 'the extent to which the development of objectives of donor governments and organisations can be adapted to or fitted into Indonesia's development objectives',[17] to both governments.

The fact that aid pledged at the IGGI was managed through formalised procedures was important, because promoters of Indonesian projects were guaranteed regular opportunities for the exposure of their proposals to an international forum and to the world's largest aid donors (an assurance not given for other recipients to nearly the same extent). Furthermore, Japanese officials knew that, because of commercial interest in using IGGI-base aid to finance development schemes, their own immediate responsibility for identifying and promoting feasible projects would be lightened.

Officials, however, should not be underestimated. The OECF, even though it was an implementing agency, had great influence with regard to Indonesia in domestic loans policy-making, because of its expertise and the extent of personal rapport built up over the years between itself and the Indonesian Government, especially with officials of Bappenas. This subject is taken up again in Chapter 8, but while career officials in the ministries changed posts regularly, OECF staffing was more stable. More importantly, systems and procedures for aiding Indonesia were soundly institutionalised, and the OECF fitted easily into the structure of loans management as a specialist loans agency.

Japan and ASEAN: Developing-country Demands

Aid management could be as easily complicated by a government's foreign policies as made smoother; bilateral aid to Indonesia through IGGI was one matter, but Japanese promises to the Association of Southeast Asian Nations (ASEAN), of which Indonesia was a member, presented a quite different set of problems. Here the concept of 'bilateral' took on new meaning. Japanese Government statements and, to all appearances, policies marked a continuation of the core of

Japanese aid policy since the early 1960s, development of the Southeast Asian economies. Japan's relations with the region were traditionally centred on economic cooperation and aid, for the purposes of 'peace, prosperity and economic security'. This underlying interest is still clearly drawn; government statements since the 1950s have made plain that economic cooperation with Southeast Asia had a security interest, in more recent years termed 'economic security'.[18]

Mr Fukuda's trip to the ASEAN countries in mid-1977 was the beginning of a new, potentially more constructive, relationship with the group. Spurred by the 'Fukuda Doctrine' (embracing the ideas of a non-military role for Japan, 'heart to heart' relations and economic cooperation on 'equal terms') the Japanese Foreign Minister, Sonoda Sunao, was active in fostering an atmosphere of earnest Japanese concern for the region, which found concrete form, following the promises of aid which Mr Fukuda gave, in apparent agreement to finance the first of ASEAN's five 'industrial projects' (the urea plant in Indonesia), assurances on ASEAN priority in Japan's ODA (seen already in the trend in bilateral loan commitments), and attempts to establish some form of commodity stabilisation agreement. There is disagreement both amongst ASEAN nations and in Japan about the benefits of such a scheme, although several commentators remind us that such questions are related to far-reaching issues of Japan's overall trade structure, its non-tariff barriers and import restrictions, and its industrial structure policies relating to direct investment overseas for offshore processing and manufacturing. The Prime Minister, Mr Ohira, also added his voice to Japan's active ASEAN diplomacy by speaking to UNCTAD V in Manila and later pledging $1 million in scholarships to ASEAN students as part of his 'people to people' diplomacy. While this may not amount to much in terms of Japan's overall relations with ASEAN, it certainly indicated a firmer government response to the need, often remarked upon, for a Southeast Asian policy.

To some extent Japan's positive response to ASEAN's development needs was prompted by persisting ASEAN demands and by criticism from ASEAN countries of dilatory Japanese responses to their requests and even to Mr Fukuda's promises of 1977.[19] The Japanese Foreign Minister was diligent in trying to promote a higher, but 'softer', Japanese profile in the region, especially in his attendances at the ASEAN Foreign Ministers' Conferences in June 1978 and July 1979. Nevertheless, the Fukuda Doctrine had its roots in the past, and its theories were those reiterated by successive Japanese Prime Ministers since Mr Kishi's visit to the region in 1957.[20] The emphasis was more on the personal contact, but the objectives did not change. Quite simply, ASEAN (and the region in which it is situated) is important to Japan and, in terms of Japan's overseas investment and trade, became more so in the 1970s. The ASEAN countries accounted for 18 per cent of total Japanese foreign direct investment in 1965 and 20 per cent in 1976. Despite the fact that Japan's interests in the ASEAN countries varied by country and by commodity, Japan depended on ASEAN for the imports of a number of important commodities, including palm oil, bananas, kerosene and heavy fuel oils, natural rubber, sawn and veneer logs and unwrought tin. This relationship was partly due to the importance of some individual members of ASEAN to Japan's economy, trade and bilateral aid flows, since all members but Singapore ranked among the top ten recipients of Japanese loans. Indonesia dominated as an aid recipient, host to Japanese investments (13.9 per cent of ASEAN accumulated total by fiscal 1977) and in trade (taking nearly half of Japan's trade with ASEAN 1975-7[21]). ASEAN has been a favoured aid recipient, however, as evidenced by the notably higher grant element on Japanese official loans which its members received: an average of 49.95 per cent between 1957 and 1976, compared with 46.83 for the Middle East (the next highest group) and 43.99 per cent overall.[22]

Here there impinged something of Japan's aid management problems, for the desire of ASEAN as a group for Japanese aid conflicted with the pressure of existing bilateral ties between Japan and each member country. Mr Fukuda's pledges to each country on his tour were certainly more substantial in form and content than those to ASEAN as a group. In addition Mr Fukuda, when promising assistance to the industrial projects, was careful to note that one condition of any aid would be 'feasibility'. Previous chapters have explained the lengthy process required to satisfy that criterion; Mr Fukuda's conditional promises were indeed restrictive, and further conditions attached to Japanese commitments to finance most of the first industrial project emphasised that the problems of implementing even prime ministerial promises were real: Japanese officials stressed that some of the Japanese funds would be tied and that the recipients would be responsible for any cost overrun due to delays in construction or inflationary pressures on costs. Nor was it clear whether the loans would be made on a strictly bilateral basis to the Indonesian Government or to what extent procurement would be untied.

The political significance of the ASEAN relationship was underlined by the fact that it had been one consistently dealt with far more publicly, and at a far more senior level, than many others. While procedures were still important, prime ministerial promises threw aid into a more political context (where, for example, the MFA's Asian Affairs Bureau would be influential in aid decisions and the Economic Cooperation Bureau might take up the reins only in implementing those promises) and focused the public gaze on the aid relationship in its widest sense. This was not necessarily bad: Japan's aid to some members of ASEAN in the past was not noted for being publicised. On the other hand, Japan's willingness to link directly aid and diplomatic objectives towards the region was seen in the move to suspend aid to Vietnam following its invasion of Kampuchea in January 1979.[23] While aid was not cut off and officials pledged to honour Japan's commitments, the indication that Japan was even considering the step was a departure from the tendency to avoid an overt use of aid in that way.[24] This indicates the extent to which control of the aid programme — even after commitment — could be taken out of the hands of the Economic Cooperation Bureau for reasons quite external to the aid system. Japan's decision to continue aid to Vietnam was later supported by the United States Government, supposedly as one means of maintaining some influence over Vietnam's allegiances.[25]

Foreign policy was therefore wedged tightly between the patterns of the past and the structures of managing the aid programme, and a highly political, sometimes acrimonious, aid relationship was not easily extracted from that context. For those reasons, a completely fresh departure in policy towards ASEAN was not possible, and the apparent hesitation by Japanese officials in implementing the vague promises of leaders was to be expected.

'Special' Relationships

The notion that certain bilateral ties were sometimes 'special' underpinned several important aspects of the aid programme, and could be used to justify individual aid agreements or arrangements.

Many pressures made some aid relationships 'special'. Obvious political and economic necessity, for example, meant that Japan's aid relations with the Middle East became very 'special' after the oil crisis of 1973. The threat of an Arab oil embargo on Japan brought a rapid response from the Japanese authorities and, after visits to the region by the Deputy Prime Minister and the Minister for International

Trade and Industry, promises of large amounts of Japanese assistance for industrial projects in the region were forthcoming. Burma, too, always had a special place in Japan's Asian policy and received the first untied loan from the Japanese Government in 1971 for oil development, a year before Japan promised to introduce LDC untying on its loans.[26] There are numerous pragmatic reasons for according countries favoured treatment; perhaps Japan reacted to not having a colonial link in the same way as Britain or France. In any case an elite group of recipients emerged, outside the normal classification of LDCs by *per capita* income. This, in turn, reinforced existing biases in Japan's aid administration (the emphasis on bilateral project loans) or created new ones, such as preferences for joint government-private financing of massive resource projects. Because not all sections of the donor administration looked on special relationships as warranted, however, inconsistencies could develop in donor policy. As a case in point, the MOF regarded aid to Indonesia as having been too freely spent and tried unsuccessfully to hold down IGGI commitments.

The Indonesian and South Korean aid relationships with Japan were 'special relationships' where economic feasibility was easily subordinated to political necessity. Aid for 'special' recipients was managed more carefully and usually by higher officials than was aid to other developing countries. The imprimatur of a bureau director or vice-minister came to be required, and both the IGGI and South Korean commitments automatically needed approval from bureau directors or ministers. Large projects in these countries were also ratified at this level, especially if (as officials repeatedly suggested) political and commercial interests were likely to be intertwined.

The Japan-South Korea Ministerial Meetings, which formed the apex of an aid relationship which was always special, were an example of how aid management could become regularised at the highest level. Aid to South Korea began in 1965 after relations with Japan were normalised, and was continued in large quantities thereafter. Following a substantial package of loans and grants that accompanied the normalisation agreement (102,000 million yen in grants over ten years for agricultural equipment, fertiliser and machinery, and 68,000 million yen in loans for economic infrastructure projects), South Korea always ranked second or third on the list of Japan's loan recipients. The ministerial meetings (held regularly only with South Korea and Australia) occurred annually and while they included discussions of topics other than aid, projected levels of Japanese commitments were announced to coincide with them. Japanese aid to South Korea was affected by the vicissitudes of the political relationship in a way unique among Japan's aid recipients. The South Koreans, Japan's closest neighbours, alone enjoyed a linguistic heritage similar to the Japanese and were able to build upon an economic base formed by lengthy Japanese occupation. The interrelationship of their countries' history, geography and culture helped forge close links between Japanese aid administrators and South Korean officials, and some Japanese officials even claimed that only the South Koreans truly understood the Japanese aid system. The South Koreans' effective use of the Japanese aid administration was a telling weapon in the bilateral aid dialogue. Here was the closest working relationship between Japanese aid and any recipient-government officials, and a clear example of the subordination of aid policy to a political relationship.

Special relationships also existed with other countries. The view that Indonesia occupied a strategic place in Southeast Asia was not confined to the MFA. Indonesia as a source of raw materials and as a suitable target for Japanese direct investment figured prominently in the minds of MITI and MOF officials. The size of Japanese loans to Indonesia for oil and natural gas development testified to the strength of these views and of political and business articulation of similar ideas. The Asahan project, for which an initial government loan was made in August 1976, was an excellent example of how private initiative and leadership encouraged government assistance necessary for overseas private investment.

Plans to develop the Asahan region in Northern Sumatra and to tap the waters of Lake Toba had a long history. The Dutch first made studies of the water resources in the area in 1908 and plans for a hydro-electricity scheme were being considered at the outbreak of the Second World War. During the Japanese occupation of Indonesia a survey of Northern Sumatra resources was partially completed by a team led by Kubota Yutaka, head of the Yalu River Hydro-Electric Power Company, and the idea of using hydro-electric power to operate an aluminium refinery was rekindled. After the war, Kubota (as President of Nippon Kōei, Japan's first firm of consultant civil engineers) tried unsuccessfully to interest both the Indonesian and Japanese Governments in Asahan as a possible reparations project, and other leading industrialists, including Iwata Yoshio, Matsunaga Yasuzaemon and Ayukawa Gisuke, attempted to initiate the scheme, but its cost proved prohibitive.[27]

It was not until 1967 that Kubota again raised the question of the Asahan project with the Indonesians and, with the backing of a Japanese Government loan, a full-scale survey was made between 1970 and 1972. Four years of slow and at times stormy negotiations followed (at one stage, Indonesia threatened to ask the Soviet Union for assistance instead of Japan), involving the Indonesian and Japanese Governments and Japanese business, and an agreement was concluded in 1975 in which five Japanese aluminium companies and the trading firms agreed to undertake Asahan development using Japanese Government finance. The 'Asahan formula' for supporting large projects underlined the complex ties existing in such schemes even when political and economic interests were in harmony. The question of donor leadership was important and Asahan negotiations were, until the final stages, directed by the group of aluminium companies chaired by Sumitomo Metal.

The companies originally disagreed with the Japanese Government when tenders were first called in 1972. Officials were not keen to support a costly dam-plus-refinery package, and the MOF was doubtful of granting a concessional loan for private investment, while the MFA was worried about the imbalances which could arise between IGGI donors. Despite these reservations, it seems that high policy – Japanese industrial relocation, support for Indonesia's political and industrial decentralisation, establishment of a Japanese presence in the Malacca Straits area – prevailed in both countries. Only when the Japanese Cabinet agreed on 4 July 1975 to designate Asahan a 'national project' (one given full government support because of its acknowledged importance in Japan's own policies) was government assistance assured. MITI officials coordinated efforts to secure government financing and the 'Asahan method' was deemed a model of effective government-business cooperation in overseas development.

Three Japanese agencies – the OECF, JICA and the Export-Import Bank – agreed to provide 70 per cent of the original cost of $880 million, while a syndicate of Japanese city banks pledged the remainder. The OECF took a 50 per cent equity share in the project consortium, P.T. Indonesia Asahan Aluminium, in partnership with twelve Japanese chemical, metals and trading companies (although some trading company representatives later admitted to being pressured by the government into participating). The OECF made an initial LDC-untied loan of 26,250 million yen (at a relatively low interest rate of 3.5 per cent over 30 years with eight years grace) in August 1976. Increased construction costs, however, placed strains on the arrangement. Inflation and design problems had by 1978 pushed the projected cost to over $1.5 billion, a figure which had both Indonesian and Japanese officials worried, especially since there were no inflation cost clauses in the original agreement. Renegotiation of the agreement was a real test

of the 'special' relationship, and Japan pledged support of the extra
burden of cost increases in August 1978.[28]

Japanese Government support for other private business activity in
Indonesia also arose directly from pursuit of official policy objectives.
The Mitsugoro maize farms established in the Central Lampung region
of Southern Sumatra were among the first efforts by the Japanese
Government and private enterprise to develop supplies of agricultural
commodities for export to Japan, but they failed as an aid project and
as an attempt to satisfy Japanese policy goals.[29] Begun in 1968, the
Mitsugoro scheme was backed strongly by the Indonesian Government
and supported financially by the Japanese authorities, who saw potential
benefits for Japanese aid, trade, and food procurement policies. By
1977, however, exports to Japan had stopped, farm production was
maintained by diversifying crop plantings away from maize, and the
company was losing hundreds of thousands of dollars per year. Nevertheless, the Japanese Government continued to provide funds for the
venture.

The Mitsugoro story depicts some basic themes in Japanese aid
policy and policy-making. The practical problems of the maize farms
were a result of haste, poor preparation and a disregard for the realities
of implementing rapidly conceived plans in an unknown region such
as Lampung. There was a basic contradiction between government
assistance to private enterprise and the goal of diversifying food import
sources on the one hand, and the development of an agricultural aid
programme on the other. While government support for private
ventures has been an essential aspect of Japanese aid policy, often
successful in building up the industrial or resources base in recipient
economies, the forces which combined in the Mitsugoro case only
revealed the inherent dilemmas.

The Mitsugoro scheme was initiated by a joint venture agreement
concluded in August 1968 between Mitsui and Co., the Japanese trading firm, and KOSGORO, a large Indonesian cooperative movement.
The project sprang from several sources. Mitsui had been eager since the
1950s to push ahead with overseas maize development, and in this
scheme received strong backing from its Indonesian counterparts and
the Indonesian government. Support from Japanese officials (especially
in the MAF) was offered out of consideration of the need to diversify
import sources and to develop import commodities overseas, and a
hurried, incomplete process of assessing the proposed site and the
project as a whole eventually led the scheme into difficulties. A leader
of the survey party, Ochiai Hideo, later wrote that the survey was too

hasty. His evaluation lists a myriad of problems that subsequently proved disastrous for the project, and is powerful testimony to the extent of original miscalculations: limited Japanese experience in and knowledge of tropical agriculture (especially that undertaken on a large scale); insufficient monitoring of the weather patterns (the seasons turned out to be irregular and the rain undependable); disease-prone maize varieties and attendant pests; unsuitable machinery and the difficulty of obtaining spare parts; the paucity of local mechanical skills; and the lack of a feeder road to the farms from the coast. Ochiai even suggested that Mitsui should have exercised more care in choosing its partners in such joint ventures.[30]

While these problems limited Mitsugoro's usefulness to Japan's food import policy, that policy itself was partly to blame. The notion of 'development and import' (*kaihatsu yunyū*) of goods needed by Japan, when coupled with government aid for such projects, set up contradictions between aid policy and domestic economic policy. This led to further official hesitation when private firms like Mitsugoro were likely beneficiaries. In this case, the Japanese Government financed the project over seven years through the OECF's 'general project' scheme of loans to Japanese overseas investors, but this did not extend to infrastructure, such as roads and bridges, essential to the marketing of the project's product. MOF and, to a lesser extent, MFA opposition meant that this was put off until 1977, well after the project's viability in providing maize to Japan was lost.

This is not to deny that the local economy benefited from the project, which is still operating on cash crops other than maize, or that Japanese officials and businessmen did not learn from their setbacks, but the separation of commercial interests from government aid policy was never fully carried through. The Mitsugoro scheme was supported by the Japanese Government for the wrong reasons and, in the end, brought few concrete benefits to Japan. Nor was its value to Indonesia conclusively proven. Although Japan was not the first country to see agricultural projects it financed succumb because of poor preparation, mistakes clearly occurred and the Japanese Government failed to act quickly enough to help rectify them. Agricultural development policy was ill formed and dominated by a 'development import' approach in which the perspective was 'from the top' and the focus on visible results and immediate returns to the donor country, neither of which was conducive to success in agricultural projects. The politics of Indonesian regional development on the one hand, and of Japanese import policies on the other, did not fit easily with the difficult process of establishing a new production and marketing organisation in an unsophisticated local economy. Confidence in Mitsui's ability to develop maize in Central Lampung was misplaced, and the rigid distinction between loans for the farms and loans for infrastructure prevented the project from being established on a firm basis from the outset.

The case has also shown that Japanese decision-making proved inadequate in a number of respects. Pressure from both Tokyo and Jakarta overrode the meagre provisions for properly assessing such projects; the simple enthusiasm of those involved eventually proved to be a poor basis for a new development venture. Domestic political structures and procedures also contributed to Mitsugoro's difficulties: the financing agency — the OECF — fell in easily with ministry and political demands, and the assortment of official sanctions required for the loans was no counter to doubtful viability, but allowed instead greater access for pressure and leverage.

The interlocking of official and private policies in foreign aid appeared also in Japanese ties with Brazil. This relationship was similar in many ways to that with Indonesia, and both were, to the Japanese, close and friendly partners in a 'special relationship'. In recent years, the economic benefits to Japan have become more obvious as Brazil's abundant natural resources are developed. Japanese Government capital assistance to Brazil stretched back to an early yen loan in 1961 under consortium arrangements, and up to September 1977 Japan lent a total of 22,839 million yen to Brazil, slightly less than the 24,500 million yen to the other large Latin American recipient, Peru. Total technical cooperation between 1954 and March 1977 was 2,700 million yen, which surpassed that to any nation in South America (except Peru, which received the same) although it was well below the leading Asian recipients. In September 1976 a far-reaching aid agreement between the two countries was signed, opening the way for greatly increased Japanese participation in Brazilian economic development. The agreement package totalled about $3,000 million, in soft loans and commercial credits, to be used for the construction of an integrated aluminium smelter and refinery, a steel plant, a harbour and paper pulp mills. These were financed by Eximbank and the OECF, while JICA provided 5,100 million yen for a 50,000 hectare maize development scheme.

The agreement was an overt response to political and economic considerations, promoted by politicians and influential bureaucrats,[31] since complementarity of the Japanese and Brazilian economies (especially in respect of resources) was well appreciated, as was the desire of private enterprise to move ahead with government-assisted investment. The same government-business cooperative pattern which emerged in the Asahan case was followed in the aluminium smelter-refinery project at Belem.

Other issues underlay the politics of the relationship. As with Mitsugoro, the MAF argued strongly for the overseas development of primary products, which culminated in JICA loans to maize development at Serad. This project was pushed by MAF International Cooperation Division officials, but was supported also by widespread bureaucratic and political pressure to activate JICA financing and, possibly, to test the JICA rules themselves which did not allow JICA loans to other than Japanese companies. The MOF was opposed, but an *ad hoc* Diet members' Study Group for the Promotion of Japan-Brazil Agricultural Cooperation under the leadership of the former Minister for Agriculture, Kuraishi Tadao, strongly advocated use of JICA funds. Although temporary, these forces certainly had an effect in translating perceived economic potential into policy. Another longer-term factor was the impact of emigration policy on Japan-Brazil relations. According to JICA, of 64,747 government-assisted emigrants who left Japan between 1952 and 1976, 51,849 (80.1 per cent) went to Brazil, and the contribution of Japanese emigration policy to Brazilian development was acknowledged in the joint communiqué between Prime Minister Miki and President Geisel in September 1976. The links between Japan and Brazil created by emigration directly affected the policy-making process, it seems, for Japanese officials claimed an affinity with their Brazilian counterparts not expressed except in other special relationships with Indonesia or South Korea. It was asserted that official negotiations were easier when certain cultural understandings could be taken for granted by both sides.[32] This argument was a serious one and demonstrated a perhaps unconscious rationalisation of an aid relationship at times strongly criticised within Japan as overly commercial. The 'cultural alliance' thesis was, in addition, a handy complement to the common Japanese theme of geographical proximity as a determinant of aid flows, or to the concept of the 'international division of aid labour'.

The Minor Politics of Aid

Bilateral aid relationships were subject also to lesser influences. 'Special relationships' did not constitute the whole of Japan's foreign aid policy; the pawns as well as the knights of the aid process profoundly affected outcomes in the bilateral context, and policy-makers accommodated the interests of many politicians, businessmen and 'hangers-on'. Policy was not simply the result of rational calculation of the economic options.

As Chapter 6 has demonstrated (and as Chapter 8 will describe further in relation to the financing agencies), the 'minor politics' of aid, the stuff of the aid process, was most evident in the preparatory stages of ordinary requests. There were few predictable patterns, but desk officers in the ministries to whom came requests from LDC governments were the first targets of pressure which was directed upwards if necessary. For a country in which aid had low political relevance, Japan had an active aid business lobby, although aid bureaucrats assiduously denied the influence, even the presence, of interest groups. Decisions, they claimed, were organisational and rational.

It is true that aid officials made the decisions, although only from the options which they perceived. Budget officers, as we have seen, were more realistic since they expected and accepted pressures and were able to balance them with the constraints of MOF policy. Aid officials, however, were reluctant to acknowledge the LDP's Special Committee on Overseas Economic Cooperation, but admitted nevertheless that they frequently attended its meetings and provided it with whatever information it required. At budgeting time also they sought its support and tried to temper its demands. This was indicative of at least some latent power which could be exercised by the committee and previous chapters have already shown the influence of the committee in the establishment of JICA and in the creation of the bond insurance scheme.

Individual members of the committee were active in aid relationships which personally interested them, as occurred in the Japan-Brazil Agricultural Cooperation group which lobbied for aid to Brazil. Even Minato Tetsurō, who has been mentioned before as a man publicly committed to the development of Japan's aid programme, had his political career to consider. Minato happened to be born in the same prefecture as Noguchi Hideo, one of Japan's most revered medical scientists, who discovered the cause of yellow fever and eventually died of the disease in Accra on the Gold Coast (now Ghana) in 1928. Minato was elected to the Second Fukushima constituency, the one in which Noguchi's birthplace was situated, in 1963 and served as its member until his death in 1977. The centenary of Noguchi's birth in 1976 led Minato to try and secure Japanese government funding of a Noguchi Memorial Research Centre in Accra.

This incident showed how a minor aid relationship expanded for reasons quite unrelated to normal development criteria. Ghana received a small amount of Japanese aid, mainly technical assistance, although some debt rescheduling was made through the Ghana Aid consortium in 1968 and 1975, and Kennedy Round food aid was donated in 1975. In that year Ghana received the fourth-largest share of Japan's technical assistance in Africa after Kenya, Tanzania and Ethiopia, and over one-third (39.2 per cent) of this was medical aid. Assistance began in 1968 after the visit of a mission to the country in 1966 under Shirahama Nikichi, an LDP member of the House of Representatives and Chairman of the Medical Aid Subcommittee of the party's Special Committee on Overseas Economic Cooperation. Until 1974, 29 experts had spent time in Ghana and 17 study missions had visited the country. The Fukushima Prefectural Medical University acted as the sponsoring organisation in Japan and Honda Kenji, Professor at the University, led the first project survey mission to Ghana in June 1968.

The concept of the Noguchi Centre originated at this time. Minato was said to have discussed the idea first with Honda; it certainly did not originate with the Ghanaian Government. It was Minato, university authorities and the Noguchi Hideo Centenary Remembrance Action Committee who lobbied the MFA, the MOF and the Ministry of Education over a period of years. Talks took place between officials of the two governments in May 1976 and, after 155 million yen was spent on a survey in 1976, a joint statement on 2 November 1976 announced the plan and the proposed construction. The MFA requested 1,000 million yen for the project as part of its grant aid request for the 1977 budget. The MOF attitude was cool but it had little room to refuse in the light of Noguchi's place in Japanese cultural history and on the occasion of his centenary. An agreement was signed in July 1977 between the Japanese and Ghanaian Governments. The proposal was well timed, well planned and matched the MOF's criteria for grant aid: it was both feasible and visible. Whether it coincided with Ghana's own priorities for aid projects was less certain.

Smooth and efficient lobbying by Minato and those associated with the Fukushima group contrasted with the aid activities of another Diet member in regard also to a minor recipient of Japan's aid. This incident demonstrated how easily bilateral aid relationships could be jeopardised, because of bureaucratic susceptibility to political pressures, inadequate procedures, or the perceptions in recipient countries of the working relationships between business and government in Japan and of the way they affected aid policy decisions.

The country involved was Papua New Guinea (PNG), then peripheral to the central interests of Japanese aid policy-makers, even though Japan was regarded as a potentially important aid donor by PNG. While aid to PNG up to 1976 was mainly given as technical assistance, a 660 million yen grant was made in November 1975 for the construction of a fisheries college, the first Japanese grant made to a South Pacific nation. Fisheries aid to that region is increasingly important to Japan's fishing interests as more states enforce their 200-mile economic zones. A bilateral agreement in December 1977 with PNG for loans on extremely favourable terms was evidence of a positive Japanese response to the articulation of consistent and firm recipient policies on the acceptance of aid (PNG, for example, refused an offer of tied grant aid). The agreement, which showed clearly how effective a potential recipient's sound understanding of the Japanese aid system and policies could be, provided for loans of 11 million Kina on a 'selective draw down' basis for projects to be chosen in accordance with PNG's national public expenditure plan.[33] A feasibility study of the Purari hydro-electricity scheme was financed by JICA from MITI funds and in 1976 was the most expensive survey then being undertaken by JICA's Mining and Industry Department. JICA was also involved in a forestry project near Madang.

The present example concerns Japanese government support, through the OECF's 'general projects' assistance scheme (the same funding provided for the Mitsugoro project in Indonesia) for Japanese private investment in PNG. Financing 'general projects' (*ippan anken*) was the original work of the OECF before it began to make direct government loans in 1965, but by 1978 financing of private Japanese companies' overseas development projects made up only about 6 per cent of total OECF overseas funding. The loan in this instance came under guidelines laid down in 1961, when an agreement with the Export-Import Bank determined that the OECF would finance projects in agriculture, forestry and marine sectors but would not be involved in plant export financing. A further agreement in July 1975, which gave the Export-Import Bank the responsibility for all funding to private enterprise, still left the 'experimental stages' of projects to the OECF.

When assessing applications from companies for funds, the OECF used four main criteria to eliminate the numerous unsatisfactory requests: (a) the firm's soundness and its experience in the kind of project proposed; (b) the likely 'public benefits' to the LDC deriving from the project (based on the country's own stated priorities, if any); (c) the MOF attitude (although except in large or special applications, approvals for overseas investment were given automatically by the

Bank of Japan on behalf of the Finance Minister), and (d) the MFA attitude. The regional bureaus of the MFA were naturally sensitive to the likely political effects of such financing, particularly if projects were on a large scale or made up a significant proportion of the flow of funds to the country in question. The OECF had a stronger voice in decisions concerning these company loans than it did on direct government loans (see Chapter 4), provided that the proposal did not breach MOF or MFA guidelines on direct investment.

In spite of the OECF's authority, it was still susceptible to external pressure, as in the case of a loan to a company named Tōkai reberā kōgyō, an engineering firm based in Nagoya with capital of 50 million yen. Founded in 1959 by its President, Matsumoto Saburō,[34] the company received a loan of 800 million yen from the OECF in early 1972 to develop oil palm on the island of New Britain in PNG. This was the company's first ever overseas development work and certainly its first oil palm venture. A joint agreement was entered into with the PNG Government in March 1972, but a dispute arose in 1975 over the design and manufacture of the oil palm mill, among other things. Even with mediation, settlement could not be reached. Despite precipitate intervention by representatives of the Japanese MFA, demanding a reconsideration of PNG's position, the PNG Government introduced legislation in August 1976 to nullify the joint venture agreement and expropriated the company's assets. An independent assessor was called upon to draw up a settlement and compensation was made to the company by the PNG Government.

While the incident revealed problems on both sides, there were serious weaknesses in Japanese decision-making. OECF officials admitted that, in retrospect, it was a mistake to have approved a substantial loan to a company totally inexperienced in oil palm development and undertaking its first overseas project on a large scale in a remote part of a country relatively unknown to the Japanese. In addition, Japanese companies had done little oil palm development and the fact that the OECF loan was tied to procurement in Japan meant that similar problems of inexperience may have arisen even with other Japanese investors. Intensive study of the proposal had been necessary, but approval was forced by the intervention of a member of the Japanese Diet. The company was registered in his electorate and he was said to maintain a financial interest in its operations, and to have had wartime associations with the New Britain area. This politician was not, however, a member of the Liberal Democratic Party but the then Secretary of the Opposition Democratic Socialist Party, Tsukamoto Saburō.

It was at Tsukamoto's insistence that the loan was originally approved and with his support that Tōkai reberā pursued its case in Japan when compromise could not be reached in 1976. Tsukamoto asked a question of Prime Minister Miki in the House of Representatives on 1 October 1976 concerning the government attitude to what he saw as a high-handed reaction by the PNG Government. This brought a response from his own party, since the statement bore no relation to party policy and was made, it considered, purely out of personal interest. Tsukamoto's involvement (and the MFA's early aggressive stance) hindered the genuine attempts by officials in the OECF and MITI to arrange agreement between the company and PNG in late 1976. Although for a short time the continuance of economic cooperation between Japan and PNG was brought into question, the OECF considered the affair to be instructive. Japanese officials were made more aware of the frailty of some government-financed operations, particularly where normal decision-making processes were replaced by pressures which diverted attention from established procedure. In PNG, the need to ensure that foreign investors clearly understood that followed investment guidelines became obvious.

David Wall's conclusion about aid relationships — that no rational criteria are used in the problem of aid allocation — can be applied equally to Japanese foreign aid. In Japan, many policy questions concerning distribution and geographical and sectoral emphasis were answered by inbuilt structural characteristics and by particular bilateral traditions and pressures. In the absence of accepted policy guidelines, bilateral pressures reinforced any tendency for policy to fragment and threw responsibility onto officials to control aid relationships at the working level. Here lay the system's plodding strength. Procedures were necessary: they ensured the implementation of aid policies and helped counteract the unpredictability of bilateral relationships; they were resilient and predictable.

Bilateral aid policy-making was open to non-bureaucratic pressures because of the multiplicity of leverage points in a dispersed aid system. In some instances these pressures worked to severely limit the options for loans (IGGI, South Korean and Brazilian aid, for example), while in others pressure was brought to bear directly on officials in responsible policy positions (Mitsugoro, Noguchi Centre, PNG oil palm development, for instance), or acted to strengthen trends in policy (Asahan).

'Special relationships' were self-serving and encouraged a concentration of aid flows on selected recipients. The essence of 'special relationships' between Japan and some recipients lay in familiarity.

Indonesia, South Korea and Brazil (among others) enjoyed favourable treatment in policy terms because of the cumulative weight on policy-makers of flows of financial, human and informational resources between Japan and those countries. The realising of Japan's economic and political aspirations, and the coincidence of recipient governments' development priorities with Japan's desire for secure and visible projects, were possible because of the ease with which 'proximate policy-makers' conversed. In regard to Indonesia, the existence of a formal aid-giving structure was helpful, but even before the IGGI began there were intimate ties between Japanese and Indonesian elites.[35] Special relationships were less susceptible to bureaucratic politics and to the influence of officials on the size and scope of aid flows. Officials were responsible for the details of policy implementation.

What each of these bilateral relationships had in common was a problem of information or intelligence. All were isolated examples, but each was in some way affected by the state of information available. The other cases examined — Asahan, Mitsugoro, the Noguchi Hospital and PNG oil palm development — can also be discussed in terms of knowledge of local conditions, history or culture. Relationships founded on precise information were more likely to be beneficial to donor and to recipient. Japanese private enterprises, especially engineering consultants whose own future depended on their success in promoting economic cooperation, acted as middle-men, often compensating for the problems which the Japanese Government had in appreciating local conditions.

Minor aid relations were particularly sensitive to the availability of precise and adequate information; the lack of such information opened aid to outside pressures so that decisions, and policy, became bound up with unstable forces. This reflected the need for strong data collection and management, or for a greater LDC input into Japanese policy-making, since the Japanese bureaucracy was more likely to act with initiative and independence when supplied with the information on recipients adequate for its procedures. It is this question which Chapter 8 takes up.

Notes

1. J. Alexander Caldwell, 'The Evolution of Japanese Economic Cooperation, 1950-1970' in Harald B. Malmgren (ed.), *Pacific Basin Development: The*

American Interests (Lexington, Lexington Books, 1972).

2. Gregory Clark, 'Japan, Australia and Asia', *Australia-Japan Relations Symposium Papers*, Canberra, 1975.

3. Nakane Chie, *Tekiō no jōken: nihonteki renzoku no shikō (Criteria for adjustment: the Japanese continuum mentality)* (Tokyo, Kōdansha, 1972).

4. Bannō Junji, 'Japan's Foreign Policy and Attitudes to the Outside World, 1945-1968' in Peter Drysdale and Kitaoji Hironobu (eds.), *Japan and Australia: Two Societies and their Interaction* (Canberra, ANU Press, 1980), and Satō Seizaburō, 'The Foundations of Modern Japanese Foreign Policy' in Robert A. Scalapino (ed.), *The Foreign Policy of Modern Japan* (Berkeley, University of California Press, 1977).

5. Bannō, 'Japan's Foreign Policy'.

6. Gaimushō jōhō bunkakyoku, *Nihon no keizai kyōryoku* (Ministry of Foreign Affairs, Public Information and Cultural Affairs Bureau, *Japan's economic cooperation*), March 1974, p. 81, suggested that geographical distance was one cause of weak relations with Africa.

7. Minato Tetsurō, 'Shiron: keizai kyōryoku hakusho: 12 kōmoku teigenshō' ('Economic cooperation white paper, a private version: a summary of my 12-point proposal'), *Kokusai kaihatsu jānaru*, 10 June 1976, p. 14.

8. Kikuchi Kiyoaki, 'Denaoshi semarareru keizai enjo' ('Economic aid under new pressures'), *Ekonomisuto*, 9 November 1976, p. 65.

9. According to a senior MFA official in an interview on 9 December 1976. Nakane argues that the Japanese lack a sense of *noblesse oblige* in *Tekiō no jōken*, pp. 159-62.

10. Gaimushō keizai kyōryokukyokuchō-hen, *Keizai kyōryoku no genkyō to tembō: namboku mondai to kaihatsu enjo* (Ministry of Foreign Affairs Economic Cooperation Bureau Director General (ed.), *Economic cooperation: present situation and prospects: the North-South problem and development assistance*) (Tokyo, Kokusai kyōryoku suishin kyōkai, 1978), Ch. 3, Part 3.

11. In 1968 the implementation of all direct loans to Indonesia was transferred to the OECF, which set up a Loan Department (now the Third) devoted entirely to Indonesian aid. The move was prompted by the need to centralise aid given through the Inter-Governmental Group on Indonesia (IGGI) and was facilitated by amendment of the OECF Law in 1968 to allow it to give commodity loans. It was first agreed upon in ministerial talks on the 1968 budget (see *Nihon keizai shimbun* (hereafter *Nikkei*), 12 January 1968).

12. George Cunningham discusses the active-reactive role debate in his *The Management of Aid Agencies: Donor structures and procedures for the administration of aid to developing countries* (London, Croom Helm, 1974), pp. 13-16. The 'active donor approach, implemented through a regular and vigorous dialogue with recipients', as suggested by Ambassador Edwin Martin, former Chairman of the Development Assistance Committee, was not accepted by the Japanese Government.

13. David Wall, *The Charity of Nations: The Political Economy of Foreign Aid* (London, Macmillan, 1973), p. 166.

14. John White, in *The Politics of Foreign Aid* (London, Bodley Head, 1974), p. 270, regards consortia as effective aid coordinating mechanisms, but Paul N. Rosenstein-Rodan regards them as only a second- or third-best solution to the problem of development financing, even though the original consortium created as an emergency measure for loans to India fulfilled its purpose 'brilliantly' ('The Consortia Technique', *International Organization*, vol. XXII, no. 1 (1968), pp. 222-30). P.J. Eldridge discusses aid to India in his *The Politics of Foreign Aid in India* (London, Weidenfeld and Nicolson, 1969).

15. Nancy Viviani, *Australia and Japan: Approaches to Development*

Assistance Policy (Canberra, Australia-Japan Economic Relations Research Project, Australian National University, 1976), p. 70. See also G.A. Posthumus, 'The Inter-Governmental Group on Indonesia', *Bulletin of Indonesian Economic Studies*, vol. VIII, no. 2 (July 1972), pp. 55-66.

16. Why the procedures changed is not altogether clear. One OECF official suggested that because the number of small projects fell and long-running projects increased, there was greater need for proper feasibility studies rather than for an identification mission. Another said that the MFA wanted to limit the influence of the OECF in decisions on projects (interviews, 17 August and 4 November 1976).

17. Posthumus, 'The Inter-Governmental Group on Indonesia', p. 66.

18. Yamamoto Tsuyoshi, *Nihon no keizai enjo (Japan's economic aid)* (Tokyo, Sanseidō, 1978), p. 195.

19. Yano Tōru, 'Japan and ASEAN', *Canberra Times*, 29 March 1979.

20. F.C. Langdon's book, *Japan's Foreign Policy* (Vancouver, University of British Columbia Press, 1973), documents this.

21. Ikema Makoto, 'The "Common Approach" to Foreign Policy, with Special Reference to ASEAN's Relations with Japan', paper to Tenth Pacific Trade and Development Conference, Canberra, March 1979, Table 1, p. 23.

22. Gaimushō, *Keizai kyōryoku no genkyō*, pp. 494-5.

23. *Asian Wall Street Journal*, 23 January 1979.

24. *Asahi shimbun*, 12 March 1979.

25. *Yomiuri shimbun*, 23 May 1979.

26. *Far Eastern Economic Review*, 11 May 1979.

27. Nishihara Masashi, *The Japanese and Sukarno's Indonesia: Tokyo-Jakarta Relations 1951-1966* (Honolulu, The University Press of Hawaii, 1976), pp. 66-71.

28. *Japan Economic Journal*, 22 August 1978 and *Nikkei*, 5 March 1978 and 3 August 1978.

29. See Alan G. Rix, 'The Mitsugoro Project: Japanese Aid Policy and Indonesia', *Pacific Affairs*, vol. 52, no. 1 (Spring 1979), pp. 42-63.

30. Ochiai Hideo, 'Indonejia Ramponshū ni okeru tōmorokoshi no kaihatsu — Mitsugoro no shichinen' ('Maize development in the Lampung region of Indonesia — seven years of Mitsugoro') in Ogura Takekazu and Yamada Noboru (eds.), *Kokusai nōgyō kyōryoku no genjō to kadai (Present situation and problems of international agricultural cooperation)* (Tokyo, Nōsei kenkyū sentā, 1976), p. 346.

31. See Ashikaga Tomomi *et al.*, 'Kokusai kyōryoku jigyōdan (3)' ('Japan International Cooperation Agency (3)') *Yunyū shokuryō kyōgikaihō*, June 1976, pp. 4-11.

32. Expressed both by officials and in an MFA publication where Central and South America were described as 'very close (*najimi no fukai*) to Japan because of Japanese emigrants'. See Gaimushō jōhō bunkakyoku, *Nihon no keizai kyōryoku*, p. 81.

33. For details, see Colleen Ryan, 'Papua New Guinea Wins Generous Aid from Japan', *National Times*, 19-24 December 1977.

34. Kōjunsha shuppankyoku, *Nihon kaisharoku (Japan company directory)*, 1976, p. 522.

35. Nishihara gives the best account of these relations in *The Japanese and Sukarno's Indonesia*.

8 POLICY AT WORK: THE CYCLE OF AID

The frontiers of the aid relationship, of the whole aid system, saw the most change and innovation. Not only consultants (although they had the most immediate reasons for developing aid projects), but varied commercial interests, politicians and others all played their part in pump-priming the aid programme. Officials tended to follow — sifting, assessing and consolidating these initiatives. Completing an aid project meant disbursing funds, 'moving money', fulfilling aid targets. It added to the donor's annual aid achievement and boosted the figures prepared by DAC every year as the recognised aid effort. But spending the funds was a job for officials; more particularly, it was the responsibility of executive agencies, not central ministries.

Seeing aid decisions through was part of the politics of foreign aid, and here officials became important in determining future options. This chapter takes up several themes of this last part of the book and discusses how the 'money' end of the aid relationship worked. In one sense, the concern is with implementation, for we examine the agencies as part of the policy process. We are not dealing with a single policy, however, as Pressman and Wildavsky did in their implementation study, yet their characterising implementation as 'a seamless web'[1] is a neat description of the Japanese aid process. While policy was chronological, there was no single, immutable policy process. Structures, ideas and modes of bureaucratic (and political) interaction were important but, as Heclo argued, 'the content of a policy can itself be a crucial independent factor in producing effects on the policy-making process . . . "new policies create new politics" . . . even the most innovative creations are decisively shaped by the content of previous policy'.[2] This chapter assesses such interaction in the aid process.

The end of a project was only a beginning. Bilateral relationships were ongoing and cumulative, a mass of criss-crossing requests, commitments and agreements about what could and could not be part of an aid programme. Consultants helped feed in a succession of new ideas and possibilities; Nippon Kōei's part in initiating and maintaining the pace of several important bilateral relationships was typical of how the foreign aid programme could grow, usually unpredictably and haphazardly. Political pressures were not restricted to single issues or policies and, just as aid budgeters saw policy in terms of past and

250

future, aid projects were an investment in future ties.

Japanese ministry officials commonly expressed the view that they made decisions which were faithfully carried out by the executive agencies. This was obviously a useful way to maintain the appearance of decision-making power, and while administrative law and bureaucratic tradition upheld this myth, bureaucratic politics effectively destroyed it, by blurring the notion of decision-*making* and broadening the concept of power in the bureaucratic setting. Decisions were made through action at the frontier of the aid relationship as much as at the centre, for approving aid was not enough; how projects and programmes were ultimately set up and made to work was the real test of the aid relationship and of the relevance of prior decisions. It was ultimately the way external-internal linkages interacted with domestic structures and procedures which shaped decisions, and influenced the effectiveness of the implementing machinery, especially agencies such as the OECF and JICA. Policy was no smooth road, no easy formula, and trying to put decisions to work affected aid in several important ways. The effective delivery of aid over the long term, while naturally dependent on the agency's own capacity, was immediately affected by the supply of information to Tokyo ministries, by the type of advice given by agencies when considering aid proposals, by the actual management of aid payments, by the lessons learned in those processes and, ultimately, by the nature of subsequent aid proposals.

The link between ongoing aid projects and a new cycle of requests and response has been established above; this also was not a simple process. One thing this book has demonstrated is the extraordinarily open structure of the Japanese bureaucracy; the leverage points in the dispersed aid system were numerous and were well used by insiders and outsiders alike. On occasions pressure was applied by one part of the system through the judicious use of external judgements — a World Bank report, DAC examinations, fellow donor criticisms or opinions. The DAC was a favourite weapon of both ministries and agencies, for its regular exhortations to Japan to improve her performance could be counted upon to contain specific barbs. Budgeting was the common setting for the bureaucratic politics of aid and it was in this context that the first round of most arguments was won or lost and the parameters of future aid laid down. The final rounds were often resolved in the way aid was actually used, yet the dimensions of those arguments were very often formed by the dynamics of spending aid budgets. External pressures from the recipient side of the relationship most consistently were channelled through the outer reaches of the aid system, although ministry contact with recipient governments or their representatives in Tokyo embassies was important too. Political dialogue at this level was essential to both parties in monitoring aid as it fitted into the broader diplomatic and foreign policy framework.

Information

Information is an essential resource of policy-makers, important (as Geiger and Hansen showed[3]) no less to aid than other areas of policy. Given that aid is not a domestic policy, information channels can become seriously complicated by the distance between the beneficiaries of policy and policy-makers. In this sense the 'frontiersmen' of Japan's aid bureaucracy were vital to the transmission of information. What was fed back to Tokyo, and how it was then used, were closely related.

Structural factors, as well as the way the agencies worked in the field, influenced the flow of information and its ultimate policy impact. The lack of professional aid officers and regional aid specialists in the ministries, particularly at more senior levels, was prolonged by the generalist traditions of the bureaucracy. Concern for career obviated the need to develop expertise in aid, least of all towards minor recipients. The absence of country programming, except for some planning of directions of technical aid for budgeting purposes, was due to lack of staff and uncertainty over responsibility for aid planning, and this inability to forecast and coordinate the aid programme for individual countries was, according to OECF officials, 'the weakest point' of Japan's aid administration.[4] As a result, the official policy emphasis on a broader distribution of aid flows was not subjected to consistent internal appraisal, nor was the concentration of aid on a few special relationships countered by a bureaucratic presence, in the form of specialist officers or programmes, representing small aid recipients. The major recipients retained their favoured positions.

This was the result not only of biases in attitudes and information-gathering but also of poor information systems within the Japanese bureaucracy. The Administrative Management Agency's 1974 investigation of economic cooperation reported that pre-feasibility surveys in LDCs were poorly prepared, surveys were unsatisfactory and the resulting assistance inappropriate. In addition, it argued that overseas embassies did not collect enough good information, and communicated badly with home ministries, so that the wrong specialists and equipment were often sent and surveys were delayed. Information systems in Japan linking ministries, agencies and research institutions were regarded as inadequate and specialists' reports and surveys were used ineffectively.[5]

The rather trenchant report by the Administrative Management Agency recommended the establishment of a more centralised information storage and retrieval system within the aid bureaucracy to enable technical reports to be fully utilised, improved embassy reporting of conditions in the developing countries, their requests and recipient-government priorities, and better methods of preparing for official surveys through the use of existing information systems. It also suggested that completed project finding and project identification surveys be used more effectively and that ministries set up consistent aid assessment procedures.

Subsequent ministry responses to the report were vague and non-committal, no doubt because the report had dealt mainly with OTCA procedures, whereas only one month after the report was released the OTCA was abolished and replaced by the Japan International Cooperation Agency. Problems continued — both JICA's annual report in 1974 and one of its senior officials recommended improved survey procedures[6] — but JICA was designed to accommodate an enlarged survey programme and the increased information management tasks which that entailed. JICA's creation did not automatically lead to a centralising of surveys; MITI continued to commission JICA surveys in the resources and mining areas, but other minor organisations (such as the Mining and Metals Agency) persisted. In addition, the OECF maintained its own survey programme and to some extent duplicated JICA's efforts. Officials put it quite bluntly that serious territorial questions were involved and, indeed, even though capital and technical aid were supposed to be efficiently meshed, competition in this respect was strong. Yet, as the two agencies learned to live with one another, methods evolved to cope with unavoidable rivalries. The OECF did not lessen its survey work (in fact, OECF 'engineering loans' extended a more permanent control over the link between project surveys and loan projects) and OECF officials took part in JICA surveys as representatives of the eventual funding agency. The OECF could formally comment on JICA survey reports, and both agencies were able formally to exchange information and documents relating to particular projects.[7]

That special provision needed to be made for such cooperation between agencies with closely integrated functions demonstrates the strength of bureaucratic divisions in the Japanese aid system. Some commentators have justified the divided and dispersed attributes of the bureaucracy on the grounds that it enabled greater access to varied information and intelligence,[8] but this claim has yet to be proven. On the contrary, it would appear that while there was certainly a vast array of relevant data available to parts of the system, it was not systematically exchanged or its collection coordinated. The Mitsugoro project, or the case of the oil palm venture in PNG are excellent examples of (among other things) the failure of intelligence. These cases pointed up the problems of making surveys quite apart from assessing their results. This is not to deny, however, that changes were being made where they were seen to be most needed.

Creating an 'intellectual infrastructure' within government circles in Japan for dealing with foreign aid was assisted by semi-government research outfits like the Institute for Developing Economies and the International Development Center. Yet work done on commission by those bodies was in theory for the commissioning agent alone and neither was closely tied into a broadly based cross-agency information network. Set up under special legislation in 1960 and eventually, after some open argument between the Ministry of Foreign Affairs and MITI over its rightful line of control,[9] responsible to the latter, the institute received about 90 per cent of its budget from MITI, but was rather independent in how it chose its research programmes. MITI officials often wondered about the usefulness of this relative independence to policy, although the institute nevertheless built up a strong reputation within Japan and overseas for its academic research on developing economies. The IDC, founded in 1971 with support from government, private and academic circles, was designed for more immediate policy purposes, notably for advising on development issues and encouraging research and discussion in Japan. It has been notably successful in stimulating some younger officials to further their interest in foreign aid and economic development issues.

The domestic debate about foreign aid became appreciably more active in the latter half of the 1970s, in response in particular to the explicit government promise to double ODA in three years and the ensuing public interest in Japan's aid programme. Yet the effect of this on the informational setting for policy was long term. Of far more practical relevance to the management of aid relationships and to the nature of policy options themselves was the role of the Japanese diplomatic mission in developing countries. The quality of information available to officials in Tokyo was vitally dependent on the mission's effectiveness as an intelligence-gatherer, although there remained difficulties in how information was used in Tokyo.

Japanese embassies were at the forefront of the bilateral aid relationship. Aid requests were channelled through the embassy and the mission's officers handled much of the negotiation leading up to an exchange of notes between governments. The embassy in any foreign service is the nucleus for intelligence-gathering networks, even though its direct effect on policy-making may be limited. Officers posted overseas by the Japanese MFA were close to events in recipient countries and desk officers in Tokyo relied on them for initial assessments of aid proposals (see Chapter 4). As Fukui states in his study of the MFA, however, there was a fine line between policy formulation and the gathering and processing of information.[10] Certainly in foreign aid the latter continually defined options in the former.

The Japanese Government was represented in over 100 developing countries, with a preponderance of manpower placed in Asian embassies. Some, such as those in Jakarta, Seoul or Manila were strongly staffed, with several officials with experience in aid policy within the ministry in Tokyo or in other missions. These and other embassies were in the past headed by senior men from MFA's Economic Cooperation Bureau such as Mikanagi Kiyohisa, Sawaki Masao and Kikuchi Kiyoaki. Yet, MFA career management policies did not normally lead to officers from economic cooperation desks being sent to the corresponding overseas mission. Taking Indonesia as an example, there were cases of senior officials moving from the level of councillor in the Economic Cooperation Bureau in Tokyo to top positions in the Japanese embassy in Jakarta, but on the whole there was no distinct pattern of movement of officers in the Jakarta post to or from positions related to Indonesian economic cooperation.[11] A period of service in the Economic Cooperation Bureau often entailed a few years in a developing-country embassy afterwards, but this did not seem to be carried through in any consistent way. The generalist traditions of the MFA worked against it.

Although the embassy was important to both the planning and implementing of foreign aid, the Administrative Management Agency's report of 1974 was critical of the embassy in the way information was collected and passed on to Tokyo, and noted that some embassy officials did not fully appreciate the nature and relevance of the aid task, so that the division of responsibility with other government agencies was ill defined. Several writers have observed intelligence failures in various areas of the MFA's work,[12] but it is difficult to generalise about what can be an extremely sensitive and constantly shifting relationship between officials of an embassy and its host government. It has been the changing function of the embassy in foreign policy-making that has most directly influenced the relevance of overseas missions to home aid divisions.

The common distinction made between the centre and periphery in Japanese organisations was explicit in the MFA.[13] Furthermore, the functions of the embassy reaffirmed the tendency of the aid administration to favour domestic, not LDC, priorities. Not being in a position demonstrably to affect the flow of aid other than in exceptional circumstances, the foreign missions reinforced prevailing patterns in aid policy-making. Duty in these posts was, of course, integral to an official's career, yet successful careers were made in Tokyo as promotion occurred within the ministry to the important geographical and (to a greater extent than in the past) functional divisions. The ease of international communication in the 1960s and 1970s restrained the embassy's role as an independent actor in the aid relationship in its own right, and it became the end point of the long arm of Tokyo bureaus and their senior officials. The embassy was formally allowed little initiative, despite the fact that the narrow data base in Tokyo left it largely up to embassy officers as to how they operated in the field.

This limited flexibility had its successes and failures. Relations between members of the Japanese mission in Jakarta, for example, and the economic agencies of the Indonesian Government such as Bappenas, the National Planning Board, were on the whole friendly and constructive.[14] Differences between donor and recipient government easily arose, however, as over the oil palm venture in PNG, which resulted in expropriation of the investing firm's assets. Such difficulties to some degree accompany all aid relationships, especially where there is mutuality and an explicit desire on both sides to secure their interests. Coordination of a mission's activities was not always helped in the Japanese case by the number of non-MFA officials stationed in overseas missions. MITI and MFA officers, for example, were often at odds in their attitude to relations between Japan and the particular LDC, and a sense of competition over the best aid proposals was frequently present.

The MFA saw no need to alter established practice to include more aid professionals or specialists in embassy staffs. Policy was still directed from Tokyo and sending Tokyo-based officers overseas on aid business as it arose was a regular feature of a diplomat's working life. The need to staff small embassies in minor recipient capitals did not warrant changes in career policies. Periodic visits from headquarters officials were seen as more effective than the posting of officers with aid experience. Even though the information-gathering functions of the overseas mission could not be fulfilled by Tokyo officials, its duties were limited both by tradition and by the manoeuvrability of ministry personnel.[15]

The embassy's participation in aid policy was therefore restricted but still necessary in the management of the relationship, in addition to its information tasks. William Wallace, echoing the 1969 Duncan Report on Overseas Representation, considered the 'buffer' function of British overseas missions to have remained valuable. 'The role of the embassy,' he wrote, 'has become much more one of establishing and maintaining contacts with the ministers and officials of foreign governments . . . of providing "the essential "door-opening" function" for "experts flown over from London for short meetings", of providing an element of continuity.'[16] In the Japanese situation, where embassies did not house the specialist staff of a central aid agency as British missions did, this function was even more valuable. Maintenance of smooth relations with recipient governments by senior and junior embassy officers partly offset the lack of aid expertise within the embassy. Negotiations carried out on orders from Tokyo were more manageable.

Overseas missions acted primarily as extensions of the central decision-making apparatus. Tokyo policy and practice reinforced this; diplomats themselves were in any case Tokyo-oriented, since it was from Tokyo they came and to Tokyo they returned, often after only two years at a post. In another sense, however, embassies were important barometers of the effects of Japan's aid policy and its future directions. Their function in the mechanics of aid decisions has been explained in earlier chapters; it was their eyes and ears at the pre-request and implementation stages of projects or aid agreements which helped fit that individual aid item into a wider relationship. Their formal duty to explain the conditions of request to developing-country governments could naturally both encourage and discourage. Just as active in this generative process were the executive agencies, the OECF and JICA, which together handled the preponderance of official development assistance.

The Agencies: Front-line or Back-room Officials?

The OECF was, in 1979, the acknowledged backbone of Japan's overseas aid programme. It provided about half of ODA, almost all official loans, was the key to the ODA-doubling policy, was extending into development surveys and innovative programmes such as lending prior to an exchange of notes, joint financing with the World Bank and extensive research on aid problems, and was closely involved in funding part of the large bank of projects agreed upon between Japan and China. At the same time, several chapters above have noted the difference between the formal exclusion of the OECF from decisions on loans and the very real influence exerted by its officers not only on immediate policy mechanisms but on the array of options open to decision-makers, the scope and type of requests and, importantly, the trends in policy as reflected in the content and direction of flows.

There are sound historical reasons for this rather late blossoming of the OECF, which had always tended to play junior partner to the Export-Import Bank, the older and larger of the two lending agencies. It was not until 1975, in fact, when the OECF was given responsibility for all ODA loans, that the commercial and development interests of Japan's overseas loan programme became more precisely disentangled.

The OECF grew out of the Asian Development Fund set up in 1957 with support from the Prime Minister, Kishi Nobusuke. Although the Development Fund was never drawn upon (its charter required other countries to contribute, which they never did), it came about at a time of growing dissatisfaction within business and government circles with the limited scope for government funding of overseas development projects. The Eximbank was criticised from several quarters as too commercially oriented, its export promotion aim being too restrictive for many development projects.[17] The rather diffuse concept of a new development financing body offering better terms than the Eximbank was brought into focus by the availability of the idle budget of the Asian Development Fund, which was used as the basis of the new OECF, formed in December 1960.

Arrangements for administrative control of the OECF posed difficulties, for it easily came within the purview of the three ministries (the MFA, the MOF and MITI) already managing Japan's aid relations. To avoid having the OECF associated too closely with any one, responsibility was given to the EPA: the OECF was placed under a domestically weak agency and at the same time officials from other ministries entered the fund in strategic policy positions. The OECF was emasculated from the start and made subordinate to bureaucratic interests rather than to those objectives of international development espoused in the OECF Law.

This control was not welcomed by all. One of the original directors of the OECF complained how little the MOF understood development ideas and how its attitude to aid was clouded by a domestic concern for budgets and financial stringency.[18] OECF personnel were, with one or two exceptions, drawn from the ministries and no new recruits entered until 1964. The General Affairs Department of the OECF was staffed by men from the MOF, the MFA and MITI because of their joint over-seeing role. The Loan Department was headed by an officer from the Bank of Japan and its two Division Directors were from the Eximbank and the Japan Development Bank, with desk-level staff transferred from the Eximbank, making the fund's Loan Department in a sense an extension of the bank. Indeed, there were early attempts to merge the two financing agencies, the first of many efforts to settle border disputes between them. The fund, in its first few years, had little opportunity for expanding and did not make its first government loan until 1965, as part of Japan's agreement with South Korea. Politicians and officials criticised the fund as being too small and argued for the division of functions between fund and bank to be made more explicit. In 1963 a committee was set up to investigate these demands, and some commentators, such as the *Asahi shimbun*, demanded a bigger role for the OECF, independent of Eximbank control, so that Japan could fulfil her obligations to the developing countries.[19] The MOF certainly did not want to see this happen, and in September 1963 proposed through the Finance Minister that the Eximbank take over the duties of the OECF, and only strong lobbying from other ministries and the President of the OECF, Yanagida Seijirō, forced the MOF to back down. Nevertheless, from the outset the ministries took control of most of the fund's affairs. The 'four-ministry deliberation system' for assess-ing loan requests (*yon shochō kyōgi taisei*) was devised, which made it difficult for the OECF as a subordinate implementing agency to assert its own identity. Under this system, the four main ministries (the MFA, the MOF, MITI and the EPA) had to agree on the provision of all loans by the fund. By the 1970s, it was the heart of the decision-making process for official loans..

The OECF of 1979 was a significantly more influential agency because loan aid for projects had become the main strand of aid policy. In legal terms, it was still an implementing agency, was supervised by a higher agency (the EPA) and decisions on where and how much it spent were made by an interministerial committee. Yet, in effect, the OECF's success in managing its loans determined the success of much of the total ODA effort. Its standing as the government's main source of concessional development finance was assured: in 1968 it was given sole responsibility for funding the Indonesian aid programme and in 1975 was allocated all funding of loans with a grant element of 25 per cent or more. JICA finance was an alternative source, but JICA could only make loans in association with the OECF or the Eximbank.

In the domestic decision-making process, OECF's formal status was that of an observer and adviser, providing briefs to ministries on the advisability of loan requests, on the disbursement position and other, usually technical, questions. In budgeting and in loans policy, however, OECF officers were actively involved. In one sense, the OECF was not well equipped for activist decision-making: its line organisation was devoted predominantly to carrying out its loans work, and staff units were few. The three operations departments were given responsibility for separate regions or countries, while the General Affairs Department and part of the Survey and Technical Appraisal Department were given general overseeing and coordination functions. The total fixed staff of the OECF numbered only 199 in July 1978, far fewer than JICA's complement of 1,056, although OECF numbers gradually increased after 1977. Despite the pattern of ministry control of senior positions in the OECF, the fund was an active recruiter of staff in its own right, and the large majority of its officers were OECF 'natives'. University graduates making a career in the OECF spent their working lives in the foreign aid area, while transferees returned to their home ministries after a few years. The fund was also old enough to have its own officers in departmental or divisional directorships. The 'aid professionals' worked in the OECF, not the ministries.

How was the OECF able to exert its influence and use its expertise when ministries were adamant that power in policy-making ought to lie with them? The fund's real power derived from its indispensable part in implementing projects: the OECF assumed the lead once the intergovernmental exchange of notes had taken place. Before a loan agreement could be signed, the OECF initiated several legal procedures and, if necessary, further studies of the proposed projects. Once the agreement was signed between the OECF and the recipient government, disbursement was in the OECF's hands. In fact, a number of hitches in the disbursement process over the mid-1970s — which occasioned sluggish loan performance — were due directly to problems at these stages of the lending process.

The structure of the OECF — mainly consisting of country divisions — meant that the fund, of any of the government bodies responsible for loans policy, had the closest contact with developing countries and their representatives. Not only that, but it was a contact continuous over the life of projects, often spanning four or five years. In some divisions, notably that devoted to looking after loans to the Indonesian

Government (the only division handling relations with just one recipient), this contact was undivided and was assisted by the presence of an OECF office in Indonesia. The OECF maintained offices in eight developing countries. To some extent OECF (and JICA) offices complemented the intelligence and reporting functions of Japanese diplomatic missions, and one OECF representative in Jakarta in the early 1970s regarded his post as that of an unofficial aid attaché. OECF offices were establised in countries with which the fund had substantial business (Thailand, South Korea, Indonesia, the Philippines, India and Egypt, for example), and the institutionalisation of that presence through the officer served to strengthen the ties at an official level, and prolonged the concentration of aid on a few recipients.

As was explained in earlier chapters, it was at the fringes of the aid relationship that the aid process derived its momentum and from where, in making policy on loans, the real aid relationships were fed. The OECF, as an executive agency, was obviously an integral part of this. As an arm of the donor government, the OECF provided a bridge to Tokyo decision-making, but was at the same time concerned to spend aid funds effectively. The OECF brought this direct country experience to bear when 'advising' ministries on loan requests. Not only did the OECF built up its own strong information base but, because it was not part of official loans decision-making, the information was used selectively.

The importance of the OECF's 'back-room' role was magnified by its influence at the political level, mainly through its President. The incumbent of this position was always a retired bureaucrat or public figure and the position itself came to be that of a type of 'aid ambassador'. The immediate past President, Okita Saburō, one of Japan's best-known international economists and proponents of a purposeful Japanese role in international aid, gave the post a professional flavour. Previous Presidents Yanagida Seijirō (formerly a senior Bank of Japan and Japan Air Lines official) and Takasugi Shinichi (a former banker and businessman) fulfilled the domestic functions of the President in budgeting and the day-to-day bureaucratic battles, as did the President after 1977, Ishihara Kaneo, successful in securing increased budgets and staff levels. As a former senior MOF official, this record is not surprising. With Okita (who in November 1979 was made Foreign Minister), however, the presidency acquired a greater 'ambassadorial' flavour. Okita had close associations with several Southeast Asian nations, and the OECF's taking over all financing of Indonesian loans in 1968 made the OECF President an influential voice in Japanese debate on aid to that country. The President was also a member of the Advisory Council on Overseas Economic Cooperation and it was he who gave evidence before the Diet if questions arose on particular bilateral relationships involving OECF loans. In 1978 the President came under the public gaze in this way over allegations of irregularities in pricing of rolling-stock for the Seoul subway, a project being financed by OECF loans.[20]

The Japan International Cooperation Agency, while less directly relevant to Japan's overall aid performance, carried an increasing burden of technical and grant aid. It was much larger than the OECF, but like that agency its senior staff were mainly ministry transferees and, given its network of 19 overseas offices and 18 other representatives, and the substantial body of expertise within the organisation itself on countries and forms of assistance (especially technical aid), its advice was essential to decisions about project finding and surveys. Its importance here bolstered the agency's position as it tried to plan for technical aid and establish its own views against the opinions of the MFA. A trend towards regional development planning by parts of JICA, and its taking over from the MFA of the important recipient contact stages of grant aid implementation (one excellent example of the pressure for change from within the aid bureaucracy) laid the foundations for an enhanced JICA say in decisions in later years.

The view that executive agencies merely effected the wishes of the ministries is too glib. The formal hierarchies of the Japanese administrative system did not apply wholesale, and agency officials certainly did not regard themselves as subordinate elements of the same system. Their actions in budgeting, in assessing aid and in managing bilateral relationships speak for themselves. The agencies were not strictly in competition with ministries, although OECF officers, for example, were quick to put forward their own opinions of particular proposals. The agencies linked several corners of the aid system: they were the donor aid representatives to the recipient countries. More so than the diplomatic mission they were bound to the actual use of aid funds at the project level over several years, and over the life of succeeding projects. In this communications role, the agencies helped integrate cross-ministry views back in Tokyo. As a source of technical knowhow and aid experience (in which the ministries were limited by staffing policies and structure), the agencies' advice underwrote the ministries' consent.

The Aid Programme: Growth and Change

There is no simple answer to the question of how Japan distributed its aid. Admittedly there were serious problems involved. Why was Japan's aid apparently so commercial in purpose? Why was it not spread more equitably across countries and regions? Was it the pawn of business and politicians? Could Japanese policy respond to criticism and the demand for change?

The policy 'implementers' were closest to the realities of the aid process. The agencies, the consultants, the project managers and assessors, the construction firms, while concerned with their immediate tasks, kept the aid programme 'on the boil'. As the permanent Japanese presence in the developing countries they were relatively free to pursue the aid relationship beyond the ongoing projects, yet together with diplomatic missions they acted, as Part III of this book has shown, within several institutional and structural constraints on bilateral aid relationships.

There were, first of all, historical and diplomatic perspectives on Japan's role as a donor, which led over the years into fairly definite patterns of geographical distribution, the emphasis within the aid programme itself on loans, and the sharing of power between parts of the aid administration. These were conservative forces, promoting gradual expansion of existing patterns of aid, and complemented by *ad hoc* political pressures — seen in several examples cited in early chapters — which sprang from basically the same perspectives on the need for mutuality in relations with the developing world. The 'special' relationship was the product of the combination of these forces.

At the Tokyo end of the aid process the stabilising influence of procedures, of budgeting and of (until very recently) a quiescent public interest, easily accepted what these constraints offered. Growth along a tried and tested path was definitely easier. It was into this policy mould that the agencies also had to fit, although managing the recipient end of the relationship inevitably meant that agency officials were exposed to more of the development process than most other Tokyo-based bureaucrats and could feed this consistently into their appraisals.

The growth and the style of Japanese aid policy stemmed initially from the lower levels of the aid administration and from beyond it. The popular conception of aid deals clinched in the private rooms of expensive Tokyo restaurants was a side issue. More important were the questions about what the deal was for, what project was involved, who carried out the survey, where the initial impetus to request came from, how it could be carried through and with what consequences. For that reason, the development survey, the energy of consultants and the agencies were the key to the growth and change of aid relationships. At that level were the options for policy most broadly defined and policy actually put into practice, and from that level appropriate information flowed into the ministries in Tokyo.

The aid process, at least in the Japanese case, can be seen as a cumulative 'cycle' of bilateral commitment along paths defined by Japanese patterns of project assessment, selection and implementation. Political imperatives in foreign aid policy acted as one of several filters — which were mainly administrative — through which the shape of the aid programme emerged. The progress of the aid relationship at the recipient's end, however, favoured several long-standing Japanese recipients because of the steady build-up of an administrative investment in Japan for managing aid to those countries. This group changed gradually over the years — the top recipients were a restricted few — but the ASEAN nations were always among them. The commercial and economic interests of Japanese representatives, both public and private, in these countries were in normal circumstances directed to maintaining and expanding Japanese aid. Until now, these interests largely corresponded with the varied political objectives of the Japanese Government, and the Japanese aid bureaucracy was not sufficiently strong nor possessed of information or expert resources to move out of this 'cycle' of predictable (and in any case safe) commitments. The result — a relatively passive approach to the donor role — was a natural bureaucratic disposition.

Administrative criteria reinforced, and at times forced, political will over the types and directions of Japanese foreign aid. This combination gave the programme its donor-oriented and conservative character and, at the same time, led to an emphasis on the quantitative aspects of the aid relationship rather than on the terms and conditions of aid given. When the donor was, through a history of close postwar relations and a strong financial, commercial and human infrastructure of Japanese investment and aid projects, tied to the recipient through political and administrative associations at all levels, the incentive to push for improvement rather than expansion in those ties was restricted.

How then has substantive change come about? Was there a point at which internal and external pressures forced responses from the bureaucracy? The aid programme was, of course, always changing, as the frontiers of the aid administration pushed ahead into new development work. Yet this effort favoured the major recipients and the setting of new directions or the acceptance of new donor functions were altogether different questions needing new solutions. International pressures helped stimulate the adoption of new ideas, and pressure came both from donors and recipients, independently and via the more reformist parts of the domestic administration. Japan's relations with her main recipients, for example, were scattered with expressions of discontent and demands by the latter for a better deal in aid, trade and investment. The January 1974 demonstrations during Prime Minister Tanaka's visit to Southeast Asia were the most violent of recent years, but continuing ASEAN complaints over Japan's delay in fulfilling the promises made by Mr Fukuda in August 1977 were indicative of constant recipient pressure on Tokyo officials.

Change, however, was predominantly incremental, since structures and procedures did not allow more than gradual shifting of the content and directions of aid flows. The relative swing away from Asia to Africa and the Middle East did not occur suddenly, but over several years in conjunction with the development of specific aid agreements and projects. Likewise, the move to support social infrastructure assistance and the 'basic human needs' programme has yet to be finally translated into firm policy commitment. Prime ministerial initiatives — such as the pledge to double aid — were similarly subject to the whims of a complex and divided bureaucracy. In the end, the direction, pace and scope of policy change were never certain, but depended ultimately on administrative responses to the politics of bilateral aid relationships.

Notes

1. Jeffrey L. Pressman and Aaron Wildavsky, *Implementation* (Berkeley, University of California Press, 1973), p. xv.

2. Hugh Heclo, *Modern Social Politics in Britain and Sweden* (New Haven, Yale University Press, 1974), p. 5.

3. Theodore Geiger and Roger D. Hansen, 'The Role of Information in Decision-making on Foreign Aid' in Raymond A. Bauer and Kenneth J. Gergen (eds.), *The Study of Policy Formation* (New York, Free Press, 1968), pp. 329-80.

4. Interview, 28 November 1978.

5. See Gyōsei kanrichō gyōseikansatsukyoku, 'Kaigai keizai kyōryoku ni kansuru gyōsei kansatsu: gijutsu kyōryoku o chūshin to shite' (Administrative Management Agency, Administrative Inspection Bureau, 'Administrative inspection of overseas economic cooperation: technical cooperation') in *Gyōsei kansatsu geppō*, no. 180 (September 1974), pp. 29-38. Ministry responses were reported in the July 1975 edition of the same journal.

6. Kokusai kyōryoku jigyōdan, *Gijutsu kyōryoku nenpō* (Japan International Cooperation Agency, *Technical cooperation yearbook*), 1974, pp. 178-9, and Tanaka Tsuneo (Director of JICA's Planning and Survey Coordination

Department), 'Kaihatsu chōsa to tekkaku na hantei' ('Development surveys and precise judgements'), *Kokusai kyōryoku*, no. 240 (September 1974), pp. 14-17.

7. Gotō Kazumi, 'Enjo gyōsei ni kansuru ikkōsatsu' ('The aid administration reconsidered'), *Ajia keizai*, vol. 20, no. 4 (April 1979), pp. 66-85.

8. Ikema Makoto, 'Japan's Economic Relations with ASEAN', paper to Tenth Pacific Trade and Development Conference, Canberra, March 1979, p. 16.

9. Cited by Chalmers Johnson, from a collection of memoirs of MITI officials, in Robert A. Scalapino (ed.), *The Foreign Policy of Modern Japan* (Berkeley, University of California Press, 1977), p. 233.

10. Fukui Haruhiro, 'Policy-making in the Japanese Foreign Ministry' in Scalapino (ed.), *The Foreign Policy of Modern Japan*, pp. 3-35.

11. Determined by cross-checking postings in Ōkurashō insatsukyoku, *Shokuinroku* (*Directory of officials*), various years.

12. For example, Nagano Nobutoshi, *Gaimushō kenkyū* (*A study of the Foreign Ministry*) (Tokyo, Simul, 1975). He mentions the 'Nixon shocks' and the oil crisis among other incidents.

13. See above, Chapter 3.

14. Interview with former senior embassy officer, 11 August 1976.

15. Japan certainly did not share the experience of the United States Agency for International Development, which maintained numerous overseas missions. The short-term nature of AID employment led to innovative, 'risk-courting' behaviour among junior officers (in contrast to the State Department), but the far flung AID structure led to 'bottomheaviness' (reliance on lower ranks for adaptive behaviour) and friction between headquarters and field. See Judith Tendler, *Inside Foreign Aid* (London, The Johns Hopkins University Press, 1975), especially Chs. 2 and 3.

16. William Wallace, *The Foreign Policy Process in Britain* (London, Royal Institute of International Affairs, 1975), p. 35.

17. This criticism came from both within government and from private enterprise. See, for example, the call by the business-oriented International Technical Cooperation Association in *Keizai kyōryoku*, no. 44 (September 1960), pp. 4-5.

18. Interview, 20 December 1976.

19. *Asahi shimbun*, 16 June 1963.

20. See reports in Japanese newspapers, January-February 1977, especially on President Okita's testimony before the House of Councillors' Budget Committee on 17 February.

CONCLUSION

The principal message of this book is that bureaucratic interests were the main determinants of the articulation of Japan's aid and economic cooperation policies. Understanding this relationship is fundamental to, first, an appreciation of the apparently dogged pattern of Japan's aid flows and, secondly, to any effective critique of the system from outside. Knowing the rigidities of bureaucratic policy-making in Japan is essential also to the recipient's view of Japanese aid, for those policies were not the child of political pressures, elite decision-making or development arguments. The structure and style of the domestic aid administration determined policy through distinct channels and long-established procedures.

Originally, confusion about the purposes of aid contributed to bureaucratic change: disputes over where the rightful 'home' for aid lay led to the rapid creation of competing administrations, and served to preserve and strengthen ministerial ideologies. This divergence of structures and perceptions spilled over into the government context. There the primacy of procedures and the dominance of the short-term perspective encouraged the view of aid as a quantity, a bureaucratic resource which had to be controlled. Aid was seen not so much as a national policy as an annually budgeted tool of policy-makers. Furthermore, the internal dynamics of the aid process, which lacked real direction or commitment, were disordered. The relationships between types of aid were tenuous and constantly changing so, although limited, budgeting was the most effective filter for the mass of items and policy emphases.

As a result, the inertia of the Japanese aid organisation tied policy into procedural routines; the predominant motive was defensive. Where responsibility was diffuse, political will inconsistent and where policies of low political import overlapped ministerial boundaries, the costs of coordination increased and policy descended into temporary compromises along lines already tried and tested. This was due also to the strains on limited manpower: creating new policies needed resources which were not available. Innovation in reforming or developing new policies was not characteristic of the system, not even of its lowest levels which, in other systems, were relied upon for innovative and adaptive behaviour.[1] The workplace environment promoted the defence

267

of territory and jurisdiction, and only with difficulty could aid questions be aired across ministries with any force. Where it did occur, as with the creation of JICA, aid became an issue only when disputes became political.

Apart from controls and procedures, another emphasis of the aid bureaucracy was projects, and at this level most innovation occurred, because of the initiative and energy of those outside the bureaucracy. Aggressive and forward-looking behaviour, by consultants and others, had two effects: it enlarged the size and scope of aid flows but tended to push policies into defined paths. By building up bilateral aid flows in restricted patterns, policy in the long-term was stifled and capital aid for projects was intensified. Therefore, the only innovative element in the aid process worked to magnify biases in policy. Bilateral relationships particularly led to hardening of the direction of aid flows. The 'aid cycle' meant that policies were themselves important in strengthening biases in the system and, in turn, constraining future aid. The system was inward-looking: implementation helped identify new policy possibilities but narrowed future policy options. This conclusion is directly relevant to questions of the mechanics of policy change and reform.

The Donor's Dilemma

The Japanese aid system had much in common with other aid donors: the intrusion of other policies and interests into aid policy, the influence of the financial authorities and of bureaucratic diffusion, all exist in Western countries to varying degrees. Although Tendler presented a compelling argument for the impact of organisation on some aspects of the performance of the United States Agency for International Development, Japan's case is distinctive: bureaucratic tradition rigidified patterns of decision, diffuse structures complicated procedures, while the precise objectives of policy were never clearly in focus. This suggests why Japanese policy appeared to be out of step with many other donors. In addition to long-standing differences with many other DAC members about how subservient aid should be to other national goals and interests (Japanese officials, it is well recognised, explicitly regarded aid as a legitimate arm of national policy, even if they could not always agree on its immediate uses), the Japanese aid system was seriously strained. The 1978 move, strongly supported at the political level, to double Japan's ODA in three years, was partly an attempt to ease pressure on the system for a change towards less restrictive policies.

Yet, while Japanese aid was strongly criticised by recipients and other donors, the policy-making process inhibited reforms. The reliance on private enterprise and other lobbyists (including, irregularly, politicians) to bridge current and future policies, ensured that the official reaction to criticism would be weak or late, or normally both. This has implications for recipients, suggesting that an active recipient stance premised on sound and clearly articulated recipient policies and an appreciation of the constraints of the Japanese system (and of how they could be eased) would encourage a more positive Japanese response. A number of important themes of this book, in fact, stand out as extremely relevant to formulating recipient approaches to Japanese policy-makers: the development of the structure and ethos of the aid machinery; the proximity of aid divisions to other policy divisions; the degree of aid specialisation in the bureaucracy; career patterns; the differential effects of budgeting on commitment and disbursement of aid; the extent of cross-ministry control of aid management; the interplay between the policy-making and policy-implementing machinery; the range of leverage points in the system; the strength of political commitment and policy advisory functions; the preoccupation of policy-makers with certain bilateral ties; methods of project assessment and implementation; the continuities in project aid administration and, importantly, the limits of the information base within the donor aid bureaucracy.

Policy-making in Japan

Two aspects of Japanese policy-making dominated this study: the vigour of bureaucratic politics of aid in Japan and the resilience of organisational processes. They were closely interwoven in the making of foreign aid policy for several reasons — notably those of Japanese administrative tradition and dispersion mentioned above — but especially because the book has dealt with continuing and fairly routine policies at the operational level. The study bears out fully the tendency in recent work to view Japanese policy-making as increasingly diverse,[2] but suggests further that, within individual policy areas, procedures dominate the policy-making process.

Foreign aid cut across the interests of a wide variety of people and institutions in Japan, yet officials held sway in an area largely removed from the central policy interests of ministries. Politics touched aid intermittently and usually in regard to isolated projects or relationships. Ministers, for instance, were regular participants in some bilateral problems but, as Chapter 2 revealed, motives were rarely consistent. Parties, too, were not normally involved in policy as parties — it was the individual Diet members who were likely to make the running on aid issues.

Private enterprise did not play the dominant role in aid policy-making in Japan. The relationships between business and official policy-makers were often interdependent, for the latter looked to business for a great deal of information-gathering and preparation in certain bilateral aid situations. While the early years of Japan's aid effort witnessed strong and commercially effective representations by companies that initiated close aid relationships between Japan and some Southeast Asian nations, in the late 1970s the 'ground rules' for policy-making were those of the bureaucracy.

Discovering how these rules were interpreted occupied much of the book. Coordination and conflict in organisations are opposite sides of the same coin. Japanese culture has always placed a high value on the containment of dissensual behaviour and the maintenance of a united front, but administrators in Japan have not always succeeded in this aim.

Formal coordination mechanisms in the Japanese aid system did not perform efficiently. The fundamental barrier to active coordination across ministerial boundaries in Japan was the identification with the primary workgroup. This pattern contrasted with horizontally stratified French organisations, or American organisations with multiple decision centres where there were complex arrangements for coordination,[3] and is of relevance to all Japanese policy-making.

Heclo and Wildavsky, in their study of British budgeting, noted three components of effective coordination: personal ties characterised by trust and confidence, constant exchange of information and ideas, and the voluntary restraint of conflicts within reasonable bounds. It is notable that this was typically the case within the primary Japanese workgroup, such as the division in a ministry. That organisational unit was, in fact, an excellent example of the ongoing coordination of goals, perceptions and activities. Despite persistent tensions, the development of intradivisional understandings fostered effective administration, provided there was strong leadership.

This form of coordination was rarely formalised beyond the workgroup unit. Physical barriers intervened, perceptual screens were altered and commitments became dislocated. There were cross-cutting loyalties, and order was governed by several factors. The size of a unit and its proximity to central policy problems enhanced the strength of its linkages to other units, whereas divisions lower in the bureau hierarchy were more concerned with their own immediate responsibilities. There was a relation here also to the grade and quality of officer participating. Type of work was relevant too, and while technical divisions may have had grounds for easy dialogue between themselves, unequal power between counterpart units could restrict communication. The fluidity of priorities in a policy area such as foreign aid made this power accounting extremely complex.

The primary group was an introverted group. The first goal was internal harmony, on which depended the resolution of external problems. One result of primary group affiliation in Japanese organisations was an apparent disinclination to work from the general to the particular. Goals and long-range strategies drew upon motivations arising from concrete benefits to the group rather than adherence to the goals *per se*. The most effective and sought-after intergroup coordination was informal, or that which was outside 'regular' and officially sanctioned channels of intergroup communication. It was not necessarily patterned as *nemawashi* (broad consultation before action is taken) normally is, and did not occur prior to action or decision. It was not 'machinery', but a cultural or habitual trait, fostered by the stress on personal ties and on order or ranking as a foundation for social relationships. The tyranny of administrative regulations in areas where policy responsibilities overlapped demanded informality. Aid procedures included unofficial consultation, although the effect was inevitably a series of short-term resolutions to problems. Dramatic shifts in aid policy or long-term aid perspectives were associated more often with consistent political pressures or other critical external influences, not with the procedural core of the policy process.

Policy Change and the Future

Foreign aid, while not a central interest of the key sources of political power in Japan, impinged none the less on the interests of several important ministries and their associated groups. It was also increasingly obvious that economic cooperation policy in the wider sense had become accepted as a permanent focus of Japan's international economic policy alongside her relations with the developed world.

Indeed, the MFA's 1978 aid report was frank in admitting to the direct relevance of aid policy to other related policies of an 'economic security' nature. As this interdependence of policy areas becomes more publicly recognised, therefore, (although, of course, the aid-trade-investment trinity was always part of MITI lore), aid policy-making will become more directly relevant to the way issues are defined and solved. This will present itself in two ways: less noticeably in what observers regard as 'aid policy' and openly in the conflicts about aid policy which have usually been effectively hidden.

The boundaries of a policy are never static: the objective conditions of policy are in constant motion, and policy itself forces changes in both the perceptions held of it and in the relationships between elements of the policy process. Structures vary and organisational routines based on them alter as power shifts (for political, budgetary or whatever reasons) within the bureaucracy. Patterns of, and participation in, policy-making reflect reactions to specific views of policy.

Policy areas are interdependent. Their boundaries overlap in a way which suggests an intricate, but constantly moving policy 'map' of government, where policy contents in one field vitally influence processes in another. There were paths and guides for this interaction in Japan's foreign aid policy. Institutional and political stability was an important environmental factor. A pervasive consensus that Japan's own economic growth demanded the alignment of aid and other policies was also a constant reference point for policy-makers, although the 'mind-sets' of the 1950s and the 1970s were decidedly different. Reactions by other donors and the international aid community were integral to the aid debate in Japan, even though policy change came only slowly.

Policy formulation involved a continuing redefinition of the broader reaches of policy, and constant attempts to sharpen the appreciation of contents and their effects. Policy was not solid and unchangeable, to be chipped away or moulded, but had instead an amoeba-like quality, continually in motion, expanding, dividing or reforming. Heclo and Wildavsky's conception of policy as 'a series of ongoing understandings'[4] marks political administrators' action over time as the arbiter of content. In Japanese foreign aid, the ideas and actions of policy-makers, the structure through which they operated, could not be divorced from decisions and the body of policy. Jurisdictional problems, for example, provided one reason for ongoing border skirmishes over the defining of aid policy. There was need also for a substantial input of information from participants beyond the bureaucracy who performed a vital function in translating their assessments of the relevance of policy into future policy. Organisational channels were the medium for change and the dynamics of policy, because the policy area was continually shifting and the seach for the best balance of perception and structure was never-ending.

As foreign aid becomes more closely identified with stated Japanese goals of properly integrating relations with the developing world into Japanese economic and foreign policies (as Prime Minister Ōhira implicitly argued in his speech to UNCTAD V in May 1979), how foreign aid policy is formulated — and, indeed, what 'foreign aid' policy comprised — will become more of a public issue. Up until the end of the 1970s it was not difficult for the conflicts arising within the predominantly bureaucratic context of policy-making to be smoothed over. Changing perceptions of policy, and change in policy itself, helped hide the fundamental differences between ministries in approaches to foreign aid. This was no longer possible as the public face of aid policy — as witnessed in a greater political commitment to foreign aid by the government — restricted the range of acceptable views on the purposes of aid, in fact hardened perceptions of aid. How this affects the policy-making process in Japan will depend on the resilience of administrative procedures and the speed with which bureaucratic interests can be aligned with political imperatives. The foreign aid experience suggests that reaction will not be as rapid as the political world, or the pace of events, might demand.

Notes

1. Judith Tendler, *Inside Foreign Aid* (London, The Johns Hopkins University Press, 1975).
2. A useful survey is T.J. Pempel (ed.), *Policymaking in Contemporary Japan* (Ithaca, Cornell University Press, 1977).
3. Michel Crozier, *The Bureaucratic Phenomenon* (Chicago, University of Chicago Press, 1964), especially Ch. 8.
4. Hugh Heclo and Aaron Wildavsky, *The Private Government of Public Money: Community and Policy inside British Politics* (Berkeley, University of California Press, 1974), p. 346.

SELECT BIBLIOGRAPHY

The reader interested in finding out more about Japanese foreign aid has several useful sources to draw on. The best statistical reference in English is the OECD publication, *Development Cooperation: Efforts and Policies of the Members of the Development Assistance Committee*, the DAC review, published annually since 1964 (although it used to come under a slightly different title). Japanese official policies are described in English in the memorandum and statistical annex prepared by the Japanese Government for the annual DAC examination of donor members; DAC comments on Japanese performance are contained in the various official DAC reports on this examination. For the reader of Japanese, both the Ministry of Foreign Affairs and the Ministry of International Trade and Industry produce official reports on aid and economic cooperation. Since 1957 MITI has issued an annual 'white paper', *Keizai kyōryoku no genjō to mondaiten* (*Economic cooperation: present situation and problems*), describing Japan's policy towards developing countries in great detail. A summary of this in English is provided in *Japan's Economic Cooperation*, published each year by JETRO. The MFA in the past devoted a section of its annual 'blue book on diplomacy', *Waga gaikō no kinkyō*, to economic cooperation and more recently published a separate report on Japanese policy and performance along the lines of the foreign policy annual. In 1978, however, the MFA's Economic Cooperation Bureau published *Keizai kyōryoku no genkyō to tembō: namboku mondai to kaihatsu enjo* (*Economic cooperation: present situation and prospects: the North-South problem and development assistance*), a large compendium of statistics and data on Japanese policy and its administration, the North-South problem in general, and international efforts at development assistance.

Using these, the reader can build up a picture of official policy, but unfortunately there are few detailed secondary works on Japanese aid and economic cooperation, let alone studies of Japan's aid administration. Some are listed below, together with English language works on aid, aid administration and policy-making in general.

Allison, Graham T. *Essence of Decision: Explaining the Cuban Missile Crisis* (Boston, Little, Brown and Company, 1971)

Ashikaga Tomomi *et al.* 'Kokusai kyōryoku jigyōdan' ('The Japan International Cooperation Agency'), *Yunyū shokuryō kyōgikaihō* (April-June 1976)

Bhagwati, Jagdish and Eckaus, Richard S. *Foreign Aid* (Harmondsworth, Penguin, 1970)

Bryant, William E. *Japanese Private Economic Diplomacy: An Analysis of Business-Government Linkages*, (New York, Praeger, 1975)

Campbell, John Creighton *Contemporary Japanese Budget Politics* (Berkeley, University of California Press, 1977)

Crozier, Michel *The Bureaucratic Phenomenon* (Chicago, University of Chicago Press, 1964)

Cunningham, George *The Management of Aid Agencies: Donor structures and procedures for the administration of aid to developing countries* (London, Croom Helm, 1974)

Daiyamondosha *Nippon kōei: kunizukuri no kishu to shite* (*Nippon Kōei: the standard-bearer for nation-building*) (Tokyo, Daiyamondosha, 1971)

Drysdale, Peter and Kitaōji Hironobu (eds.) *Japan and Australia: Two Societies and their Interaction* (Canberra, Australian National University Press, 1980)

Fujiwara Hirotatsu *Kanryō no kōzō* (*The structure of the bureaucracy*) (Tokyo, Kōdansha, 1974)

Fukui Haruhiro *Party in Power: The Japanese Liberal Democrats and Policy-making* (Canberra, Australian National University Press, 1970)

Gaimushō kanshū *Me de miru gijutsu kyōryoku* (Ministry of Foreign Affairs, *Technical cooperation illustrated*) (Tokyo, Kokusai kyōryoku suishin kyōkai, 1977)

—— *Me de miru namboku mondai* (*The North-South problem illustrated*) (Tokyo, Kokusai kyōryoku suishin kyōkai, 1976)

Gotō Kazumi, 'Enjo gyōsei no ikkōsatsu' ('The aid administration reconsidered'), *Ajia keizai*, vol. 20, no. 4 (April 1979), pp. 66-85.

Halliday, Jon and McCormack, Gavan *Japanese Imperialism Today* (Harmondsworth, Penguin, 1973)

Harari, Ehud 'Japanese Politics of Advice in Comparative Perspective: A Framework for Analysis and a Case Study', *Public Policy*, vol. XXII, no. 4 (Fall 1974), pp. 537-77

Hart, Judith *Aid and Liberation: A Socialist Study of Aid Policies* (London, Gollancz, 1973)

Hasegawa Sukehiro *Japanese Foreign Aid: Policy and Practice* (New York, Praeger, 1975)

Heclo, Hugh *Modern Social Politics in Britain and Sweden: From Relief to Income Maintenance* (New Haven, Yale University Press, 1974)

—— and Wildavsky, Aaron *The Private Government of Public Money: Community and Policy inside British Politics* (Berkeley, University of California Press, 1974)

Honda Yasuharu *Nihon neokanryōron* (*A study of Japan's new bureaucracy*) (2 vols., Tokyo, Kōdansha, 1974)

Huang Po-wen Jr *The Asian Development Bank: Diplomacy and Development in Asia* (New York, Vantage Press, 1975)

Iida Tsuneo *Enjo suru kuni sareru kuni* (*Aiding and aided nations*) (Tokyo, Nihon keizai shimbunsha, 1974)

Johnson, Chalmers 'Japan: Who Governs? An Essay on Official Bureaucracy', *Journal of Japanese Studies*, vol. 2, no. 1 (August 1975), pp. 1-28

—— *Japan's Public Policy Companies* (Washington, DC, AEI-Hoover, 1978)

Kanryō kikō kenkyūkai (ed.) *Ōkurashō zankoku monogatari* (*Horror stories of the Finance Ministry*) (Tokyo, Yell Books, 1976)

Kokusai kyōryoku jigyōdan *Kokusai kyōryoku jigyōdan nempō* (*Japan International Cooperation Agency yearbook*) (Tokyo, annual)

Kōno Kazuyuki *Yosan seido* (*The budget system*) (Tokyo, Gakuyō shobō, 1975)

Langdon, F.C. *Japan's Foreign Policy* (Vancouver, University of British Columbia Press, 1973)

Loutfi, Martha F. *The Net Cost of Japanese Foreign Aid* (New York, Praeger, 1973)

Malmgren, Harald B. *Pacific Basin Development: The American Interests* (Lexington, Mass., Lexington Books, 1972)

Minato Tetsurō *Nihon no ikiru michi: kokusai kyōryoku no kihon seisaku o mezashite* (*Japan's lifeline: towards basic policy for international cooperation*) (Tokyo, Kokusai kyōryoku kenkyūkai, 1975)

—— 'Shiron keizai kyōryoku hakusho: 12 kōmoku teigenshō' ('Economic cooperation white paper, a private version: a summary of my twelve point proposal'), *Kokusai kaihatsu jānaru*, 10 June 1976, pp. 1-17

Nagano Nobutoshi *Gaimushō kenkyū* (*A study of the Foreign Ministry*) (Tokyo, Simul, 1975)

Nagatsuka Riichi *Kubota Yutaka 1966* (Tokyo, Denki jōhōsha, 1966)

Nakane Chie *Japanese Society* (Berkeley, University of California Press, 1970)

—— *Tekiō no jōken: nihonteki renzoku no shikō* (*Criteria for adjustment: the Japanese continuum mentality*) (Tokyo, Kōdansha, 1972)

Nawa Tarō *Tsūsanshō* (*MITI*) (Tokyo, Kyōikusha, 1974)

Nihon no kanryō kenkyūkai (ed.) *Oyakunin sōjūhō* (*How to use bureaucrats*) (Tokyo, Nihon keizai shimbunsha, 1971)

Nishihara Masashi *The Japanese and Sukarno's Indonesia: Tokyo-Jakarta Relations 1951-1966* (Honolulu, The University Press of Hawaii, 1976)

Ogawa Kunihiko *Kuroi keizai kyōryoku* (*Black economic cooperation*) (Tokyo, Shakai shimpō, 1974)

Ogura Takekazu and Yamada Noboru (eds.) *Kokusai nōgyō kyōryoku no genjō to kadai* (*Present situation and problems of international agricultural cooperation*) (Tokyo, Nōsei kenkyū sentā, 1976)

Ōkita Saburō *Japan in the World Economy* (Tokyo, Japan Foundation, 1975)

Olson, Lawrence *Japan in Postwar Asia* (London, Pall Mall Press, 1970)

Organisation for Economic Cooperation and Development, Development Assistance Directorate *The Management of Development Assistance: A summary of DAC countries' current practices* (Paris, 21 August 1975)

Pempel, T.J. 'The Bureaucratization of Policymaking in Postwar Japan', *American Journal of Political Science*, vol. XVIII, no. 4 (November 1974), pp. 647-64

—— (ed.) *Policymaking in Contemporary Japan* (Ithaca, Cornell University Press, 1977)

Rix, Alan G. 'The Future of Japanese Foreign Aid', *Australian Outlook*, vol. 31, no. 3, pp. 418-38

—— 'The Mitsugoro Project: Japanese Aid Policy and Indonesia', *Pacific Affairs*, vol. 52, no. 1 (Spring 1979), pp. 42-63

Rondinelli, Dennis A. 'International assistance policy and development project administration: the impact of imperious rationality', *International Organization*, vol. 30, no. 4 (Autumn 1976), pp. 573-605

Rubin, Seymour *The Conscience of the Rich Nations: The Development Assistance Committee and the Common Aid Effort* (New York, Harper and Row, 1966)

Scalapino, Robert A. *The Foreign Policy of Modern Japan* (Berkeley, University of California Press, 1977)

Simon, Herbert A. 'Birth of an Organization: The Economic Cooperation

Administration', *Public Administration Review*, vol. XIII (Autumn 1953), pp. 227-36

Suzuki Yukio *Keizai kanryō: shinsangyō kokka no purodūsā* (*Economic bureaucrats: producers of the new industrial state*) (Tokyo, Nihon keizai shimbunsha, 1969)

Taniuchi Makoto *et al. Gendai gyōsei to kanryōsei* (*Contemporary administration and the bureaucracy*) (Tokyo, Tōkyō daigaku shuppankai, 1974)

Tendler, Judith *Inside Foreign Aid* (London, The Johns Hopkins University Press, 1975)

Tsuji Kiyoaki *Gyōseigaku kōza* (*Studies in Administration*) (Tokyo, Tōkyō daigaku shuppankai, 1976), especially vols. 2, 3, 4 and 5

Viviani, Nancy 'Problems of aid administration and policy formulation among Western countries' (unpublished paper, Canberra, Australian National University, 1977)

Vogel, Ezra F. *Modern Japanese Organization and Decision-making* (Berkeley, University of California Press, 1975)

Wall, David *The Charity of Nations: The Political Economy of Foreign Aid* (London, Macmillan, 1973)

Wallace, William *The Foreign Policy Process in Britain* (London, Royal Institute of International Affairs, 1975)

Ward, Robert E. *Japan's Political System*, 2nd edn (Englewood Cliffs, Prentice-Hall, 1978)

Watanabe Takeshi *Ajia kaigin sōsai nikki* (*Diary of an ADB President*) (Tokyo, Nihon keizai shimbunsha, 1973)

White, John *Japanese Aid* (London, Overseas Development Institute, 1964)
—— *The Politics of Foreign Aid* (London, Bodley Head, 1974)

Yamaguchi Jinshū *Konsarutanto dokuhon* (*A consultant's reader*) (Kokusai kaihatsu jānaru, 1976)

Yamamoto Masao *Keizai kanryō no jittai: seisaku kettei no mekanizumu* (*The real economic bureaucracy: the policy-making mechanism*) (Tokyo, Mainichi shimbunsha, 1972)

Yamamoto Tsuyoshi *Nihon no keizai enjo* (*Japan's economic aid*) (Tokyo, Sanseido, 1978)

Zaisei chōsakai *Kuni no yosan* (*The nation's budget*) (Tokyo, Dōyū shobō, annual)

INDEX

Administrative change 49-51
  see also Japan, aid administration
Administrative Management Agency
  (AMA) and JICA establishment
    proposal 54
  manpower reduction policy 94
  report on technical cooperation
    252-3
  'scrap and build' policy 58-63
Advisory Council on Overseas
  Economic Cooperation 51, 102-6,
    208
  creation 102
  functions 102-3
  influence of officials 103-5
  lack of promoter 103
  meetings 102, 104
  recommendations 104-6, 208
Africa
  and Japanese aid 128, 223, 226
  and Nippon Kōei 204-5
African Development Fund
  and loss of recognition of aid
    128, 226
Agriculture, Forestry and Fisheries,
  Ministry of (MAFF, formerly
  MAF or Ministry of Agriculture
  and Forestry)
  aid staff 97
  choosing consultants 214-15
  four-ministry committee, non-
    membership 89, 140-1
  interest in aid 89
  International Cooperation
    Division 52-3
  JICA establishment and 52-77
    passim; control of JICA 67;
    internal ministry debate 56;
    JICA personnel 73-4; lobby-
    ing 57; proposal 52-8
  loans 140-1
  role in aid 53
  technical assistance 133-5, 164-6
Aid *see* Foreign aid
Aid budget
  and aid policy 150, 181-3
  and grants 122-5, 163
  and JICA establishment 52-62

and loans 142-3, 166-9
and multilateral aid 127-8, 163-4
and surveys 198-9
and technical assistance 133-8,
  164-6
appeals negotiations 179-81
budget timetable 160-1
carryover 184n7
Fiscal Investment and Loan
  Programme 151-7, 183-4n5
General Account 151-7, 183-4n5
growth 155-7
incrementalism 172-4, 182-3
lobbying 175-7
ministry shares 157-9
planning 138, 161, 184n10
requests 160-3
size of requests 169-74
supplementary budgets 183n4
  see also Coordination
Aid consortia 248n14
  and Japan 228
Aid donors
  administration 83-5; lack of
    power base 84
  advisory bodies 101-2
  budgeting 150, 183n1
  control of aid 50
  major decisions 146n1
  policy reform 268-9
  policy-making 15-16
  political control 83-5
  specialism 100
  see also Development Assistance
    Committee, Japanese aid
    administration *and entries
    for individual donor countries*
amakudari
  and JICA 67-8, 71-2, 73-4
Asahan project 236-8
  and industrial relocation 42
  origins 201, 203
  see also Indonesia, Kubota Yutaka
Asia
  Japanese aid to 36, 221-8
  Japanese loans to 226-7
  Japanese perceptions 221-2
  OECF loans to 227

278

see also Japan, aid donor role
Asian Development Bank (ADB) 33,
  126-9, 226
  and Japan's multilateral aid 126-9
  Japan's role in 128-9
  see also Japan, aid, multilateral aid
Association of Southeast
  Asian Nations (ASEAN)
  and Japan 231-4
  as important recipient 44-5
  industrial projects 141
  Japanese aid to 222-3
  see also Fukuda Takeo
Australian Development Assistance
  Bureau (ADAB) 101-2

Bond insurance scheme 176-7,
  186n36
Brazil
  early loans to 25
  Japanese aid to 240-1
  Japanese emigration to 241
Budget Bureau *see* Finance, Ministry
  of, Budget Bureau
Budgeting *see* Aid budget
Bureaucracy *see* Japan, bureaucracy
Bureaucratic politics 16, 58
Burma
  and Nippon Kōei 201-2
  reparations to 201-2
  special relationship 235

Cabinet
  role in aid policy 109-10
  see also Ministers, Prime Minister
Conference on International
  Economic Cooperation (CIEC)
  130
Consultants
  and ministries 214-15
  and policy-making process 191-2,
    200
  choosing 210-15
  completing projects 215-17
  definition 218n7
  government assistance to 207-8,
    211
  industry 200, 205-6
  nature of firms 205-6
  role in aid relationships 216-17
  selection, biases in 215-16
  see also Development surveys,
    Engineering Consulting Firms
    Association, Nippon Kōei
Coordination 270-1

in budgeting 175, 181-3
  see also Aid budget, Japan, aid,
    coordination
Corruption
  Opposition allegations of 111-12

Democratic Socialist Party (DSP)
  245-6
  and aid 111
  and JICA 70
  see also Political parties
Developing Countries, MFA view of
  90
Development Assistance Committee
  (DAC) 28-31, 130-2
  and Japanese aid administration
    130-2
  annual aid review 12
  definition of ODA 14
  examination of Japan 132
  influence on Japan 130-2
  Japanese membership 28-30, 130
  Japanese objectives 28-30, 130-1
  Japanese role in 131
  Japanese view of 29, 130
  US influence 29-30
  see also Aid donors, Japan, aid,
    multilateral aid
Development Assistance Group
  (DAG) 28-30
  becomes DAC 28
  Japan joins 28-9
  US influence 29
  see also Development Assistance
    Committee
Development import 78n10, 239-40
  and JICA 53-4, 56, 70-2, 78n10
  MITI support for 26
Development surveys
  choice of consultant 210-15;
    connection to project 213;
    experience 211-12; informa-
    tion 212-13; JICA attitude
    210-11; lobbying 213-14
  geographical concentration 193
  improvements needed 209-10
  JICA and 192-3, 193-200
  phases 209
  role in aid policy 191, 193
  types 192-3
  see also Consultants, Nippon
    Kōei, Private enterprise,
    Projects
Diet
  and aid budgeting 151

control of aid 108
JICA debate 70-2
ministries' attitude to 109
no aid committee 109
role in aid policy 108-9
role of individual members in aid
111-13, 168
see also Cabinet, Ministers,
Political parties, Prime
Minister
Diplomatic missions *see* Japan,
diplomatic missions

Economic cooperation *see* Japan, aid
Economic cooperation minister
and JICA establishment 59-64
and MFA 62-4
suggested by Tanaka 59
why suggested 63-4
Economic Planning Agency (EPA)
89-90
aid administration, role in 89-90
and JICA establishment: agency
proposal 55; JICA bill 69
and OECF budget 166-8
early ideas about aid 23-35
loans, role in 141
OECF, responsibility for 258
weakness of 90
Engineering Consulting Firms Association (ECFA) 199, 205-9
establishment 206
in Africa 204-5
in choice of consultants 214-15
lobbying 206-7
membership 205
source of advice to government
206, 208
surveys 207
see also Consultants, Development surveys, Projects
Export-Import Bank of Japan
(Eximbank)
and JICA 66, 69
and OECF 41, 139, 148-9n34,
259
budget 151, 179-80
established 22

Finance, Ministry of (MOF)
aid administration, role in 89
aid budgets 151-83 *passim, see
also* Aid budget
Budget Bureau: budget timetable 159-60; Economic

Cooperation Desk 169;
examiners 174-5; International
Finance Bureau and 167-8;
loans, attitude to 142-3;
multilateral assistance and
127-8; power of 159-60, 181-
2; requests to 159-69, 169-72;
response to requests 174-5;
technical assistance and 133
draft budget 177-81
EPA, influence over 90
grants, role in 122-5, 163
growth in aid, contribution to
33, 37-8
harsh attitude to aid 37-8, 40,
42-3
influence through budgets 181-3
International Finance Bureau
92; loans, role in 142-3,
167-8; multilateral aid, role
in 127-32
JICA establishment, role in
budget policy 57-8; JICA
personnel and MOF 73-4;
JICA proposal, view of 53
loans, role in 141-3, 166-8
multilateral aid, role in 126-32,
163-4; internal conflicts over
127-8, 132
technical assistance, role in 133-7,
164-5
Foreign Affairs, Ministry of (MFA)
Accounting Division 161, 166,
168-9
budget: grants 163; planning 138,
161, 184n10; technical
assistance 164-6
careers in 99-100, 255
developing countries, view of 90
DAC, view of 29, 131-2, *see also*
Development Assistance
Committee
diplomatic missions 139-40, 255;
see also Japan, diplomatic
missions
early ideas of aid 23-36; *see also*
Japan, aid, economic cooperation
economic cooperation administration 35, 36
Economic Cooperation Bureau
36, 87; grant aid, role in
122-5; importance 90-1; loans,
role in 140-3; multilateral aid,
role in 126-32; Policy Division

and budgeting 176; size of
divisions 94, 96-7; technical
assistance, role in 134
grant aid, role in 122-5
JICA establishment, role in 54-77
passim; *amakudari* 73-4; drafts
bill 64; economic cooperation
minister 63-4; lobbying 57;
OTCA 62-3; outcome 75; own
proposal 54-6, 61
loans, role in 140-3
multilateral aid, role in 126-32,
163-4; inter-bureau conflict
129-30
regional bureaus 123, 140
technical assistance, role in 133-7;
budget 164-6; MFA and MITI
135-6
Foreign aid
allocation of 228, 246-7
bureaucratic politics of 16
conflicting pressures in 83
definitions of 14, 47n21
political importance of 13
see also Aid donors
Four-ministry committee *see* Japan,
aid administration
Fukuda Takeo 11, 41, 106, 232-3,
265
and aid budget increases 181
and JICA establishment 57-62
Fukuda Doctrine 41, 232-3
see also Association of Southeast
Asian Nations, Japan, aid
doubling plan

Ghana
and Japanese aid 242-3
and Nippon Kōei 205
Grant aid *see* Japan, aid, grants
Great Britain
Overseas Development Ministry
lack of power 85
see also Aid donors

Hirai Michirō 65-9
Hōgen Shinsaku 73-5, 79-80n38
Hori Shigeo 57, 61, 63

Ikeda Hayato 29, 37
Implementation 250-2
see also Japan, aid, implementa-
tion, Japan International
Cooperation Agency, Overseas
Economic Cooperation Fund

Income Doubling Plan 1960 27
Incrementalism
in budgeting 172-4, 182-3
see also Aid budget
India 25
Indonesia 235-40
dam projects 203-4
Inter-Governmental Group on
Indonesia 228-31
LNG and aid 112
Nippon Kōei in 203-4
reparations 111-12, 203-4
see also Asahan project, Inter-
Governmental Group on
Indonesia, Mitsugoro project,
Nippon Kōei
Industrial Structure Council 103
Information 247, 252-7
and consultants 211-17
and coordination 144-6
role of 252-7
see also Japan, diplomatic
missions
Inoue Shirō 129
Inter-Governmental Group on
Indonesia (IGGI) 33, 228-31
and loans 141
International Development Association (IDA) Japanese contribution
to 164
International Trade and Industry,
Ministry of (MITI)
aid ideas 23-37 *passim*, 39-40, 42,
see also Japan, aid, economic
cooperation
bond insurance scheme and 177,
186n36
consultants, assistance to 207-8,
211
consultants, choice of 214
DAC, view of 29
economic cooperation administration 35, 36
Economic Cooperation Department 87-9, 91-2
economic cooperation report 23,
24, 26, 30, 37, 40, 42
Economic Planning Agency,
influence over 90
JICA establishment and: bill 67;
lobbying 57; outcome 75;
personnel 73-4; proposal 53-8
technical assistance, role in 133-7;
administration 134-5; budget
164-6; MFA and MITI 135-6

Iran, and surveys 212
Ishihara Kaneo 261

Japan, aid
    agencies, role of 262, *see also*
        Japan International Coopera-
        tion Agency, Overseas
        Economic Cooperation Fund
    aid lobby, lack of 85
    aid targets 39
    and Africa 128, 223, 226
    and Asia 33, 36, 221-8
    and Association of Southeast
        Asian Nations 44, 222-3,
        231-4
    and trade 23-4, 26, 35, 37,
        39-40, 42
    and Vietnam 25, 123-4, 234
    budget sources 151-5
    budgeting 176-7, *see also* Aid
        budget
    bureaucratic change 49-51, *see
        also* Japan, bureaucracy
    consultants 191-2, 216-7, *see also*
        Consultants
    coordination 144-6
    country programming, lack of
        252
    criticism of 11-12
    cycle of aid 264-5
    definition 14, 23-38, *see also*
        Japan, aid, economic
        cooperation
    disbursement:
        of grants 125, 147n13; of
        loans 144
    domestic groups and budgeting
        85, 176-7, 199
    donor role 222, 248n12, 264;
        active vs reactive role 135,
        264; in Asia 222; kudos from
        126, 128, 223, 226; natural
        role 27; perceptions of 222;
        see also Development Assis-
        tance Committee
    economic cooperation: and DAC
        30-1; and Economic Planning
        Agency 23; and Ministry of
        Finance 37-8, 40; and Ministry
        of Foreign Affairs 24, 26-7,
        36, 41-2; and Ministry of
        International Trade and
        Industry 23-4, 26, 37, 39-40,
        42; no government view 35
    future 271-3

geographical distribution 33-4,
    38-9, 41, 44, 221-8, 252
Ghana 242-3
grants 118-25 *passim*; administra-
    tion 119-25; budget for 122-5,
    163; disbursement of 125,
    147n13; implementation 124;
    lack of political pressure 125;
    purpose 118-19; rationale 119;
    recipients 119, 124
ideas about 14-15
implementation 250-65 *passim*;
    factors in 251-2; information
    and 252-7; Japanese diplo-
    matic missions 254-7; JICA
    role 133-4, 262; OECF role
    257-62; surveys and 252-3
Indonesia 235-40; Asahan project
    236-8; IGGI 228-31;
    Mitsugoro project 238-40
inflexibility of budgeting for 183
information, role of 252-7
international division of aid
    labour 222
Japanese diplomatic missions and,
    135-6, 139-40, 254-7
legislation, lack of 87
loans 25, 43, 138-44, 166-9,
    226-8; and budget 166-9;
    basis of aid policy 138; nature
    of 138, 143; principles 138-9;
    priorities 142
major recipients 225, 227-8
minor politics of 241-6
motives 15
multilateral aid 43, 125-32;
    administration 126-32; and
    budget 163-4; Asian Develop-
    ment Bank 128-9; benefits to
    Japanese policy 126; effort
    126; recipients 126; types
    125-6; *see also* Development
    Assistance Committee
Papua New Guinea 141, 243-6
periods 22; 1950s 22-5; 1958-61
    25-6; 1960s 31-3; 1969-73
    38-9; after 1973 40-4
policy change 263-5
policy priorities and budgets
    155, 157
political support for 51
Prime Minister's role 103
process 267-8
projects 191, 226
recipients *see entries for*

individual countries
source material 274
South Pacific 244
special relationships 234-41;
  Brazil 240-1; Burma 235;
  Indonesia 235, 236-40; Middle
  East 234-5; South Korea
  235-6
studies of 45
surveys 191
technical assistance 133-8; and
  budgets 164-6; criteria 135-6;
  effort 133; planning 138;
  policy 135; recipients 136-7;
  sectors 148n29
terms 33, 39, 44
United Nations, aid to 125-6,
  129-30
untying 39, 147n3
visibility, need for 223, 226
Japan, aid administration 83-113
  passim
  advisory bodies 101-7
  and grant aid 122-5
  and loans 138-44
  and multilateral aid 125-32
  and policy-making 267-9
  and technical assistance 132-8
  and UNCTAD 129-30
  central aid agency 50-1
  coordination 127, 144-6
  Development Assistance Commit-
    tee, relation to, 130-2
  'equal partnership' in 89, 118,
    139
  four-ministry committee 36, 89,
    139-44; procedures of 141-3
  growth 35-6, 50-1, 92-7
  initiative in 98, 101
  ministries 87-92
  policy advice 101-7
  procedures 118-46 *passim*
  size and growth 92-7
  specialist agencies 100, 257-62
  specialist officials, lack of 91,
    99-100, 252, 256
  structure 85-97
  see also Advisory Council on
    Overseas Economic Coopera-
    tion, Aid budget, Japan,
    bureaucracy *and entries for
    individual ministries and
    agencies*
Japan, aid doubling plan 11, 21, 27,
  153, 162, 268-9

amounts required 43
grant aid in 122
Japan, Asian roots 222
Japan, budget
  Fiscal Investment and Loan
    Programme 151-7
  General Account 151-7
  role in policy 150, 181-3
  see also Aid budget
Japan, bureaucracy 16, 83-101
  career vs non-career staff 99-100
  centre vs periphery 91, 256
  civil service examinations 99-
    100, 115n22 and 25
  collective action 98
  coordination 98, 144-6
  diplomatic missions 135-6, 139-
    40, 257
  division (*ka*), role of 98-9, 101
  generalism, problems of 100
  implementation 251
  information 252-7
  policy change 263-5
  restrained growth 94
  role in society 87, 97-8
  scope for rivalry within 100-1
  see also Administrative change,
    Japan, aid administration *and
    entries for individual
    ministries*
Japan, diplomatic missions 135-6,
  139-40, 254-7
Japan, food imports 52-4
Japan, foreign policy *see* Foreign
  Affairs, Ministry of, Japan, aid,
  and Asia *and entries for individual
  countries*
Japan, investment
  1950s 22, 25
  1960s 31
  1969-73 38-9
  1973-78 40
Japan, policy-making 15-16, 251,
  269-71
  see also Japan, aid administration,
    Japan, bureaucracy, Policy-
    making
Japan and the DAC 28-31, 130-2
  effect on Japan 30-1
  MITI view 29
  reasons for joining DAG 28-9
  US influence 29
  see also Development Assistance
    Committee
Japan Committee for Economic

Development 51
Japan Communist Party (JCP) and
    JICA 70
    and Seoul subway scandal 112
Japan Emigration Service (JEMIS) 55
    abolished 59-63
Japan International Cooperation
    Agency (JICA)
    and Brazil 241
    budget 164-6, 185n14 and 15,
        198-9
    control of 67-8, 164-6
    establishment 49-80 *passim*;
        administrative change 49-51;
        and food imports 52-4; and
        Overseas Technical Coopera-
        tion Agency 55-6, 62-3;
        benefits to ministries 75;
        budgeting 57-8, 59-62; bureau-
        cratic conflict 52-8; Cabinet
        reshuffle 56-7; central aid
        agency 50-1; Diet 70-2; draft-
        ing legislation 64-9; economic
        cooperation minister plan
        60-4; Economic Planning
        Agency attitude 55; MAF vs
        MITI 52-8; MFA attitude
        55-6; Minato Tetsurō 55,
        58-9; ministers 59-62; MOF
        view 53; OECF, relation to
        66-9; personnel 73-4; politi-
        cians 54-7; problems of bill
        65-9, 70-2, 75-6; public
        appraisal 74-6; relevance to
        policy-making 76-7
    financing 66-9, 75-6, 165, 185n15
    grant aid, role in 124
    implementation, role in 262
    initiative 138
    legislation 64-72
    OECF, relation to 66-9, 75-6
    surveys 66, 192-200; administra-
        tion of 198; budget 198-9;
        choice of consultant 210-15;
        criteria for 199-200; MITI role
        199; staff, lack of 200
    technical assistance 133-8; budget
        136-8, 164-6; MFA control
        134-6; MITI influence 134-6
    see also Japan, aid, implementa-
        tion, Japan, aid technical
        assistance
Japan Overseas Development
    Corporation 55
Japan Socialist Party (JSP)

and aid 111-12
    Indonesia reparations and 111-12
    JICA, view of 70
    see also Political parties

Kikuchi Kiyoaki 39
Kimura Toshio 223
Kishi Nobuske
    Asian Development Bank and 128
    Asian Development Fund and
        OECF 128, 258
    Indonesian reparations and 112
    Southeast Asian diplomacy 25,
        233
Kōmeitō, view of JICA 70
Kubota Yutaka 201-5
    and Asahan project 236-8
    see also Nippon Kōei
Kuraishi Tadao 54, 57

Laos, and Nippon Kōei 202-3
Legislative control *see* Diet
Liberal Democratic Party (LDP)
    aid budgeting, role in 160, 177,
        181
    factions 112-13; and JICA
        establishment 57, 58-64, 77,
        80n43
    Policy Affairs Research Council
        54, 68, 160, 168
    Special Committee on Overseas
        Economic Cooperation 50-1,
        110-11, 242, 243; Agricultural
        sub-committee 54-5; budget,
        role in 168, 177, 181; JICA
        establishment, role in 54-5,
        64
    see also Political parties
Loans *see* Japan, aid, loans

Miki Takeo 36, 50
Minato Tetsurō 41, 55, 58-9
    and aid to Ghana 242-3
    budget lobbying 176
    Ministerial Committee 106
    on Japan's Asian role 222
Ministerial Committee for the
    Economic Development of South-
    east Asia 33
Ministerial Committee on Overseas
    Economic Cooperation 106-7
    see also Advisory Council on
        Overseas Economic Coopera-
        tion
Ministers, role of 270

and JICA establishment 59-62
Ministries, Accounting Division
director 162, 168-9
Mitsugoro project 238-40
Mitsui and Company 238-40
Mongolia, grant to 119, 147n8, 163
Moriyama Shingo 55, 92
Multilateral aid *see* Japan, aid, multi-
lateral aid

Nagano Shigeo 102
Nepal, and Nippon Kōei 203
Nippon Kōei Company 200, 201-5,
211-12
experience 211
in Africa 204-5
in Burma 201-2
in Laos 202
in Indonesia 203-4, 236-8
in Nepal 203
in Vietnam 202
origins 203
reparations, benefits of 202-4
see also Asahan project, Consultants, Engineering Consulting
Firms Association, Kubota
Yutaka
Noguchi Hideo 242
see also Ghana, Minato Tetsurō

Official Development Assistance,
definition of 47n21
Ōhira Masayoshi 11, 21, 59-64, 273
Ōkita Saburō 102, 261-2
Opposition parties
aid attitude to 111-13
criticism of government 111-12
JICA bill and 70-2
see also Diet, Political parties *and
entries for individual parties*
Organisation for Economic Cooperation and Development (OECD) 14
Japan's membership 28, 30
origins of DAC 28
see also Development Assistance
Committee
Overseas Economic Cooperation
Fund (OECF) 257-62
budgets 151, 153, 155, 166-9,
179-80, 185n5
control of 258-9
Eximbank and 41, 139,
148-9n34, 259
general projects 244-5
IGGI aid 231

implementation, problems of
143-4
implementation, role in 260-1
in Indonesia 248n11
JICA, relation to 66-9
loans, decision-making for 143-4
loans to Asia 227
ministries and 259
Mitsugoro project, role in 238-40
origins 35, 258
President's role 261-2
role in aid 259-60
structure 260-1
surveys, role in 208, 219n31
untying and 39
see also Economic Planning
Agency, Japan, aid,
implementation
Overseas Technical Cooperation
Agency (OTCA)
established 36
no agricultural aid given 52
problems of 75
why abolished 62-3
see also Japan International
Cooperation Agency

Pakistan 25
Papua New Guinea 141, 243-6
Paraguay 25
Policy
definition of 18n10
growth and change 263-5, 271-3
nature of 272-3
Policy-making 15-16, 269-71
budget, influence of 150, 181-3
coordination 144-6
donors 15-16
implementation 133-4, 144,
250-1
in Japan 12-16
JICA, relation to 76-7
loans 143-4
policy change and 269
procedures for 269
processes of 12-13
see also Aid donors, Japan, aid
administration, Japan, bureaucracy, Japan, policy-making
Political parties
and aid policy 110-13
individual members and aid
111-13
factions and aid 112-13
role of 270

see also Diet, *entries for individual parties*
Prime Minister and aid policy, 103,
110
see also Cabinet, Ministers
Prime Minister's Office, Councillors'
Office 96, 103-4
Private enterprise
consultants 216-17
projects and 209-14, 215-17
role of 270
surveys and 199
see also Consultants, Engineering
Consulting Firms Association,
Nippon Kōei, Projects
Projects
choice of consultant 209-15
consultants and 207
final stages of 215-17
IGGI, role in 229-30
implementation of 250-1
problems of 200
project-finding 213
role in aid policy 191, 226
see also Consultants, Development surveys, Private
enterprise

Reparations 22, 24, 200-4
budget for 151
Burma 201-2
consultants and 200
Indonesia 203-4
Nippon Kōei, role of 201-4
Vietnam 202

Sawaki Masao 29
South Korea 106, 235-6
and Nippon Kōei 204
South Pacific 244
South Vietnam
grants to 123-4, 147n9

Japanese aid to 234
loans to 25
Nippon Kōei and 202
Southeast Asia
as markets for Japan 23
necessary to Japanese security 44
stability and prosperity argument
27
Surveys see Development surveys

Takasugi Shinichi 261
Tanaka Kakuei 265
JICA establishment, decision
about 59-64
reshuffles Cabinet 57
Tōkai reberā kōgyō 245-6
Tokonami Tokuji 54-5
Tsukamoto Saburō 245-6

United Nations, and Japanese aid
125-6, 129-30
United Nations Conference on Trade
and Development (UNCTAD)
36-7
and Japanese aid administration
129-30
UNCTAD I 36-7
UNCTAD II and III 39
UNCTAD V 11, 21, 273
United States Agency for International Development (AID)
centre vs periphery 91
no domestic power 15, 85
see also Aid donors, Development
Assistance Committee

Vietnam see South Vietnam

Watanabe Takeshi 129

Yanagida Seijirō 261
Yoshida Tarōichi 129